Dinosaurs
and Creation

Dinosaurs
and Creation

QUESTIONS AND ANSWERS

DONALD B. DEYOUNG

A Division of Baker Book House Co
Grand Rapids, Michigan 49516

Published by Baker Books
a division of Baker Book House Company
P.O. Box 6287, Grand Rapids, MI 49516-6287

Printed in the United States of America

Library of Congress Cataloging-in-Publication Data

DeYoung, Donald B.
    Dinosaurs and creation : questions and answers / Donald B. DeYoung.
        p.   cm.
    Includes bibliographical references and indexes.
    ISBN 0-8010-6306-X (pbk.)
      1. Dinosaurs—Miscellanea. 2. Creationism—Miscellanea.
    I. Title.

QE861.95 .D4 2000
231.7'652—dc21
                                       00-031280

For current information about all releases from Baker Book House, visit our web site:

              http://www.bakerbooks.com

Contents

Part 1   Dinosaurs and History

Part 2   Dinosaurs and Modern Times

Contents

Part 3   The Dinosaur Family

Part 4   The Biology of Dinosaurs

Part 5   The Physics of Dinosaurs

List of Figures
and Tables

Note: References are to question numbers rather than page numbers.

Preface

Dinosaurs hold a special fascination for nearly everyone regardless of age. These large animals inhabited a mysterious past. Unfortunately, most dinosaur books appear to *breathe* evolution on almost every page. The origin, dominance, and death of dinosaurs are all placed in the distant evolutionary past. Furthermore, many experts say that dinosaurs are the ancestors of modern birds. This book takes a totally different approach. The dinosaurs are seen as part of the supernatural, six-day creation week. Their superb design shows intelligent planning, not random genetic variation. In addition, the dinosaurs did not live hundreds of millions of years ago, but instead walked on the earth in the not-so-distant past.

For readers steeped in evolutionary history, the ideas in this book may seem incomprehensible. My challenge to such readers is to at least consider the creation possibility, as extreme as it may sound initially. What if, in the end, the creation story is correct and naturalistic evolution is wrong? The implications are many, and are as profound as life itself.

For readers sympathetic to creation, this book provides a refreshing alternative to the secular theories about dinosaurs. It is a fascinating challenge to fit these impressive creatures into literal biblical history. This book's occasional references to evolutionary change and to timescales of millions or billions of years are included simply to show how evolutionists look at history. However, the view supported through-out this book is that of a recent, literal six-day creation. This acceptance of supernatural creation is not outdated or dis-proven, nor is it an antiscientific position. Instead, creation is a serious challenge to the excesses of secular science.

There are several excellent creationist dinosaur books aimed mainly at a youthful audience. I have listed sources for these books in the back reference sections. However, an interest in dinosaurs stays with us as the years go by. There-fore, this book is written for advanced students and older readers. It includes some of the technical details of dinosaur study. The deeper we explore, the richer the dinosaur story becomes. There is no consensus on whether dinosaur names should be capitalized or italicized. I have done so for par-ticular species *(Stegosaurus)* but not for suborders (stego-saurs). A special thanks to Jorie Bail for preparing the art-work in the figures.

Many unanswered questions about dinosaurs remain for both creationist and secular scientists. I have tried to hon-estly lay out these uncertainties. But there is much that we *do* know about dinosaurs, and these facts greatly strengthen the creation position.

Introduction

Many readers have visited Dinosaur National Monument or other museums of natural history. On display are the impressively large bones from the dinosaur world. Often these bones are carefully assembled to show the largest animals ever to walk upon the earth. Signs accompanying the museum displays tell us several things:

Dinosaurs long ago evolved from even older creatures called *thecodonts* or *paleotetrapods*.

Dinosaurs lived on the earth for nearly 200 million years. However, this incredibly long time span covers only about 4 percent of earth's total history.

Dinosaurs vanished about 65 million years ago in a great extinction event. This loss occurred long before mankind evolved.

Evolutionary descendants of dinosaurs still live today, including the birds.

For the literal six-day creationist there is a basic problem with these statements: None of them are acceptable. Instead, in the creation view:

Dinosaurs appeared supernaturally during days five and six of the creation week. This week also included the origin of all other kinds of plants and animals, and culminated with our first parents, Adam and Eve.

Dinosaurs lived together with mankind during early earth history just thousands of years ago, not millions of years.

Dinosaur representatives survived the global flood of Noah's day on board the Ark. Then, during the postflood centuries, a cooler climate probably led to a gradual dinosaur demise, along with the demise of many other creatures and plants.

Living plants and animals on the earth today are only a small remnant of the original, perfectly created kinds.

Clearly the creationist view of dinosaurs presented in this book is at odds with the interpretation of most natural history museums. More specifically, what does the creation worldview entail? The following four points are not exhaustive, but they illustrate the creationist position.

First is a belief in a literal creation week. The biblical days of creation are not taken as ages, nor are gaps of time inserted between the days. Instead, the miraculous creation week was much like our present week with 24-hour days. God could have created everything in 6 microseconds, or over 6 trillion years. However, He chose six days as a calendar pattern for us (Exod. 20:11). As John Calvin said in 1554, "God himself took six days, for the purpose of accommodating his works to the capacity of men." The laws of science were also established during this first week. Therefore, current science simply cannot explain the supernatural origin details. By definition, miraculous events are beyond scientific understanding.

Second, the completed universe was made mature and fully functioning. This particular view rules out the theory of a gradual big bang expansion of the universe. It also rules out macroscopic evolution.

The third point calls for a short timescale for the earth and the entire universe beyond. Popular creationist estimates are between 6,000 and 10,000 years of total history. Obviously, creationists have alternate interpretations for fossils, radiometric dating, and stellar evolution.

Fourth, the global Genesis flood is taken as a pivotal point in biblical and geologic history. Creationists interpret much of earth's sedimentary rock strata in terms of this worldwide flood, which took place several thousand years ago.

As one might expect, there are many unanswered questions within the creation model. This is true of any scientific endeavor. However, the creation view also provides many new avenues to explore. Regarding dinosaurs, for example, interesting new predictions arise regarding their origin, diet, design, and extinction. Creation also sheds light on some of the current dinosaur controversies: Were these creatures cold- or warm-blooded? Were dinosaurs intelligent or dull-witted? Were they fast moving or sluggish? Such ideas will be explored in the following pages.

Part One
Dinosaurs and History

1. What is the evolutionary chronology of dinosaurs?

To answer this question, the conventional evolutionary timescale will be used. Geologists divide earth history into four major *eras* of time. These eras are in turn subdivided into shorter periods, some of which are listed in table 1. The Precambrian era is said to comprise 86 percent of earth history, a time span during which there was little if any life on the earth. Following the Paleozoic era of early life, the Mesozoic era is said to be the age of reptiles, including the dinosaurs. Dinosaurs first appear during the Triassic period, assumed to begin about 245 million years ago. Dinosaur ancestors are thought possibly to be the thecodonts, dinosaur-like Triassic animals. The origin of the thecodonts themselves remains a mystery in this view.

In evolutionary terms, the dinosaurs were very successful animals. The dinosaur world existed from the Triassic period through the Cretaceous period, ending about 65 million years ago. (The 1993 movie *Jurassic Park* was named for this particular time in geologic history.) The dinosaurs' existence on the earth extended roughly from 245 to 65 million years ago, or 180 million years total. Some researchers extend this time span to 200 million years, a longer span than any other

Table 1

The geologic eras, periods, and their time spans

Era	Period	Millions of years ago
Cenozoic ("Recent life")	Quaternary	1.6–present
	Tertiary	66.4–1.6
Mesozoic ("Middle life")	Cretaceous	144–66.4
	Jurassic	208–144
	Triassic	245–208
Paleozoic ("Ancient life")	Permian	286–245
	Carboniferous	360–286
	Devonian	408–360
	Silurian	438–408
	Ordovician	505–438
	Cambrian	570–505
Precambrian		4600–570

The precise boundaries of evolutionary time are not well defined. Dates shown are taken from the Geological Society of America, as of 1998. The dinosaur age is shown by the shaded region.

creature before or since. Therefore, it is commonly said that dinosaurs were the most successful animals ever to inhabit the earth. In contrast, modern humans are said to have appeared within only the past 100,000 years. In this view, the dinosaurs lived over 1,000 times longer than mankind. Even so, under the assumption of a 4.6-billion-year-old earth, dinosaurs flourished during just 4 percent of earth's entire history. A popular idea today is that the dinosaurs somehow evolved into modern birds. In this case the dinosaurs did not become extinct after all, but instead still exist today on our birdfeeders (see Question 9).

Table 2 illustrates the immense evolutionary timescale of earth history. This table's dates compare all of earth history to a single year. In conflict with the creation view, the table shows mankind appearing only in very recent earth history. Secular paleontologists are greatly disturbed by pictures that show dinosaur species and people living together. Their worldview requires a 65-million-year span between dinosaurs and mankind. Likewise, paleontologists do not believe that all dinosaurs shared the earth at the same time. For example, the stegosaurs and allosaurs are thought to precede the tyrannosaurs by about 64 million years, the same evolutionary time span that separates us from the tyrannosaurs.

The one resource in geology that is commonly assumed to be almost limitless is time itself. In a 1954 issue of *Scientific American,* Nobel Prize winner George Wald wrote about time in relation to the origin of life:

> Given so much time,
> The "impossible" becomes possible,
> The possible probable
> And the probable virtually certain.
> One only has to wait.
> Time itself performs the miracles.

The creation view challenges this assumption of a nearly endless timescale. God is the miracle worker in nature, not time and chance. Furthermore, even if *trillions* of years were available, the spontaneous origin and evolution of life would remain impossible. Both probability and thermodynamics totally rule out the whole idea.

17

Table 2

A calendar chronology of geologic time

Date	Event
January 1	Planet Earth forms (4.6 billion years ago).
February 10	Single-celled organisms originate in the sea.
July 25	Photosynthesis begins.
November 15	Oxygen gradually builds up in the atmosphere.
November 20	Precambrian era ends, Paleozoic era begins. Vertebrates first appear.
December 13	Mesozoic era begins.
December 15	Dinosaurs first appear. Birds appear. Appalachian Mountains form. Continental drift occurs.
December 20	Flowering plants arise.
December 26	Dinosaurs disappear. Rocky Mountains form.
December 31	Mankind appears.

Evolutionary earth history is compressed into a single calendar year. Note that mankind is thought to have evolved in very recent earth history.

Scientific American reprinted George Wald's article 25 years later, in 1979, in *Life: Origin and Evolution*. A revealing note was added:

Although stimulating, this article probably represents one of the very few times in his professional life when Wald has been wrong. ... Harold Morowitz, in his book *Energy Flow and Biology*, computed that merely to create a bacteria would require more time than the Universe might ever see if chance combinations of its molecules were the only driving force.

Creationists heartily agree with this rare disclaimer by *Scientific American.*

---

2. Where do dinosaurs fit into biblical history?

In the literal biblical view, the dinosaurs did not become extinct millions of years before mankind appeared. Instead, all plants and animals were part of the original creation week. On the fifth day "God created the great creatures of the sea and every living and moving thing with which the water teems ... and every winged bird according to its kind" (Gen. 1:21). These creatures would have included fish, sea mammals such as whales, marine reptiles, birds, and flying reptiles. All the remaining land animals appeared on the very next day when God made the wild animals, the livestock, and the creatures that move along the ground, each according to their kind (Gen. 1:25). These land animals included the dinosaurs. Mankind also appeared on this sixth day.

In past years some well-meaning people have suggested that dinosaurs never really existed. Fossil bones were said to be an evolutionary hoax, or perhaps they were tricks of Satan to mislead people regarding origins. However, dinosaurs did indeed once live on the earth in many places. Their bones and footprints are found almost everywhere. Just considering the letter *A,* they are found in Africa, Antarctica, the Arctic, Argentina, Arizona, and Australia.

Some have attempted to place dinosaurs in a pre-Adamic world that supposedly existed long before the creation account of Genesis 1. It is further proposed that a former created civilization lived during this early time. Divine judgment upon this pre-Genesis world is then said to be responsible for the earth's fossils and the sedimentary rock layers.

This view is called the Gap Theory, with an indeterminable amount of time elapsing between Genesis 1:1 and 1:2. However, scriptural and physical evidence for such a former world is lacking. The idea seems to be an unnecessary complication to biblical history.

It has also been suggested that dinosaurs were genetic mutants of the animal world, or perhaps they were evil monsters that arose following the Genesis 3 curse upon nature. However, dinosaurs were part of the original creation week, and certainly were not genetic misfits. They were created with purpose and beauty, and had their proper place in the early world.

The creationist view as presented here is opposite to that of evolutionary thinking, where dinosaurs and mankind are separated by 65 million years. Instead, as part of the creation week, Adam named the dinosaur types along with all the other animal kinds. Rather than the modern dinosaur labels, the names chosen by Adam were probably closer to Old Testament words such as *behemoth* and *leviathan*. The early interaction between mankind and dinosaurs can only be imagined. It is even possible that some dinosaur types were domesticated. It is only after the great flood of Noah's time that the instinctive fear of mankind was placed within animals (Gen. 9:2). Surveys indicate that half of U.S. adults believe that humans and dinosaurs at one time coexisted in spite of continued efforts by evolutionary biologists to dispel this popular idea.

At the time of the Genesis flood, dinosaurs still lived on earth. Smaller, youthful representatives were preserved aboard the Ark during the yearlong flood. After the flood these creatures and other animals left the Ark and began to repopulate the earth. Following the flood, however, a cold glacial period gradually developed worldwide. Animals that could survive the colder climate, or perhaps could migrate to warmer equatorial latitudes, survived this postflood ice age. Ice accumulation in the vast continental glaciers caused ocean levels to drop by hundreds of feet worldwide. This opened up vast land bridges hundreds of miles wide whereby animals could move across the earth between the continents. Dinosaurs, however, probably faced gradual extinction during this time due to the severe cold. Note, however, that some dinosaurs still existed at the time of the patriarch Job in the postflood world (see Question 14).

The former land bridges and continental coastlines, now submerged, might eventually yield much information to archaeologists about life on the early earth. For example, the Bering Strait is a 100-mile-wide sea separation between Siberia and Alaska. Exposed at an earlier time, this bridge may have allowed people, dinosaurs, and other animals to move from Asia to North America during early postflood centuries.

----

3. Can dinosaur fossils be dated radiometrically?

Radiometric dating does not apply directly to fossils. Typical minerals that comprise fossils—quartz or calcite for example—do not contain the needed radioactive isotopes. Likewise, the sedimentary rock layers that hold the fossils are not usually dated radiometrically. This is because sedimentary material consists of preexisting rock fragments. Minerals extracted from granite or basalt are the materials of choice for radiometric dating. Occasionally such igneous material is found within sedimentary layers, which can then be assigned an age. However, many creation scientists are skeptical of the reliability of these ages.

Radiometric dating involves the measurement of "parent" and "daughter" atoms within a rock sample. Figure 1 illustrates the process with an hourglass. The daughter atoms result from the gradual disintegration of radioactive parent atoms. Both kinds of atoms are commonly called *isotopes*. Very precise chemistry is needed since these isotopes often exist in rocks only in parts per million or parts per billion. Table 3 lists the popular isotopes commonly used in rock dating. *Half-life* is the time required for 50 percent of a group of parent atoms to decay. The large half-life values show that these parent atoms are quite stable and decay very slowly over time.

Figure 1

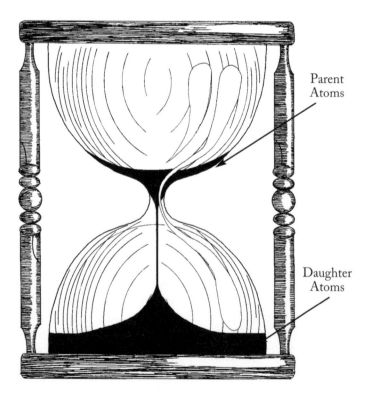

Parent
Atoms

Daughter
Atoms

An illustration of radiometric decay. The hourglass represents the rock, and the sand grains are the parent and daughter isotopes.

Radiometric dating is the only method available for the absolute dating of rock, that is, giving rock an actual age in years. Therefore, an independent cross-check on the result is missing. Experiments do indeed detect radioactive decay products within rocks, including strontium, argon, and lead. It is usually assumed that these daughter atoms formed within the rocks very slowly from the radioactive parent atoms. However, radioactive decay may have been greatly accelerated in the past, which would reduce a rock's true age. Such a rapid, *apparent* aging may have occurred at the

Table 3

Radioactive isotopes commonly used in radiometric dating

Parent Isotope	Daughter Isotope	Half-Life (billions of years)
Potassium-40	Argon-40	1.25
Rubidium-87	Strontium-87	48.8
Samarium-147	Neodymium-143	106.0
Rhenium-187	Osmium-187	43.0
Thorium-232	Lead-208	14.1
Uranium-235	Lead-207	0.704
Uranium-238	Lead-206	4.47
Carbon-14	Nitrogen-14	5,370 years

Parent atoms, daughter atoms, and approximate parent half-lives are listed. The last entry, carbon-14, is in a separate category, and is used for dating organic material rather than rocks.

time of creation, the curse, or the flood. Parent or daughter atoms may also have migrated *into* or *out of* a rock sample over time. After all, a rock is not a *closed system* like the hourglass illustration. Instead, a rock is an *open system* that is sensitive to its surroundings. Whatever the case, radioactive measurements are open to various interpretations. Radiometric dating has not proven with certainty that rocks are ancient. Many scientists, whether or not they are creationist in their thinking, remain skeptical of an overreliance on the radiometric dating method.

Note that carbon-14 dating, the last entry in table 3, does not apply directly to this discussion. Radiometric dating requires igneous rock that has formed directly from magma. Carbon dating is not used with rocks or fossils, but instead is limited to once-living things that incorporate carbon, including vegetation and shell fragments. Radiometric rock dating has been with us for a century; the carbon-14 method was discovered in the 1950s. The carbon technique has found useful application in archaeology. As long as the method is not carried too far back in time, it does not necessarily conflict with the creation view. Carbon-14 typically gives dates of thousands of years, not millions or billions.

---

4. What was the world like at the time of the dinosaurs?

Dinosaurs lived in a world far different from today. Continental drift had not yet carved out our present land masses. In the creation view the separation of continents probably occurred rapidly during the year of the Genesis flood, when the earth's entire crust was fractured and faulted (Gen. 7:11). In contrast, evolutionary thinking suggests continental drift is the result of gradual seafloor spreading over 20 million years during the Mesozoic era.

In the preflood world the Rockies, Alps, and Andes Mountain ranges did not yet exist. There was abundant tropical vegetation and numerous wetlands and lakes existed. There probably were no deserts, no major ice caps, and no severe seasonal changes. Instead, from creation until the flood, the entire earth enjoyed an ideal climate, somewhat like the "Garden of Eden" in nature. There also may have been an invisible *vapor canopy* surrounding the earth from the time of Adam to Noah (figure 2). This high-altitude layer of water vapor is an optional feature of the creation view. Such a canopy would have warmed the entire earth like greenhouse windows in the sky. Its eventual collapse could have supplied some of the floodwaters in Noah's day.

During the preflood years, the polar areas of the earth experienced the normal long periods of darkness. These result from the earth's tilt, whereby little or no sun reaches the Arctic and Antarctic regions during their winter seasons. With the mild preflood climate, dinosaurs probably lived in these remote regions. These hardy animals may have adapted to the somewhat cooler temperatures far from the equator.

Figure 2

An early vapor canopy of moisture may have surrounded the earth. This moisture was in the upper atmosphere, 12–18 miles high. Notice that the continental landforms were different in the preflood world.

_____

5. How does evolution explain dinosaur extinction?

Scientists have presented dozens of dinosaur extinction theories. Ideas range from humorous to the truly bizarre, including these:

The earth's climate became either too warm, cold, dry, or wet for dinosaur health.

A nearby supernova, an exploding star, flooded the earth with intense radiation, resulting in fatal mutations.

A "death star" called Nemesis periodically approaches the solar system and wipes out living creatures by raining comets down upon the earth.

A nearby collision between two neutron stars bombarded the earth with deadly muon particle radiation.

There were deadly giant meteor showers or meteor storms.

A passing comet poisoned the earth's atmosphere with chemicals, perhaps cyanide.

Earth's oxygen levels decreased and the dinosaurs suffocated.

The male dinosaurs grew too large to successfully reproduce.

Too many offspring were born of one gender.

Hormonal disorders arose in adult dinosaurs.

Overpopulation.

Extreme hay fever plagued the dinosaurs.

Mass suicide was carried out by dinosaur herds.

Dinosaurs were weakened by slipped vertebral discs.

Dinosaurs slowly accumulated genetic disorders.

A laxative plant in dinosaur diets disappeared, and they died of constipation.

A ratlike mammal evolved that ate the dinosaur eggs.

Weakened eggshells developed.

New poisonous plants evolved and were eaten by dinosaurs, perhaps poisonous mushrooms.

Dinosaurs ate plants containing drugs and died of an overdose.

Viral infections swept the planet.

Dinosaurs simply could not compete with evolving mammals.

Either starvation or overeating occurred.

Deadly insects evolved that carried parasites and disease.

Monster storms called hypercanes temporarily destroyed the earth's protective ozone layer.

One particular extinction theory has gained wide acceptance since the 1980s. It involves the impact of a large meteorite, asteroid, or comet, about 10 kilometers (6 miles) in diameter. The collision with earth is said to have occurred about 65 million years ago at the end of the Cretaceous period. The large object is estimated to have been traveling more than 50 times faster than a jet aircraft at the time of impact. This would have resulted in the dissipation of a tremendous amount of kinetic energy. There resulted worldwide firestorms, tsunami or tidal waves, and severe storms with acid rain. The scenario further pictures great clouds of dust, soot, and smoke resulting from the impact explosion. This material was carried completely around the earth by jet stream winds, and sunlight was thus blocked for many months or years. Plants could not grow in the semidarkness, including the vast numbers of sea plankton. The dinosaurs gradually starved from the resulting collapse of food chains worldwide. Altogether, 70–99 percent of the earth's living species are said to have gone extinct around the time of this catastrophe.

A possible collision site has been identified at the northern end of Mexico's Yucatan Peninsula. Called Chicxulub (pronounced as CHEEK-shoe-lube) after a nearby town, there is a craterlike circular formation 100–200 miles in diameter buried beneath a kilometer of water and rock. This geologic formation conveniently has been dated at 65 million years old. As further evidence of a collision, at various earth locations a clay layer at the Cretaceous-Tertiary (K-T) boundary appears to hold an unusual amount of the element iridium. The letter *K* is used for Cretaceous instead of *C*, from the German spelling of the word. Iridium atoms in the 2-inch-thick clay layer occur at only about 9 parts per billion, but this is still 30 times higher than in the surrounding rock. This particular element is known to be somewhat more abundant in meteorites, asteroids, and comets than on the earth. Thus the iridium atoms support the idea of dust fallout from an extraterrestrial impact. The collision idea was first popularized by physics Nobel laureate Luis Alvarez and his geologist son Walter who discovered iridium in the K-T clay layers of Italy in 1980. Since then, iridium has also been found at K-T boundaries in Denmark, New Zealand, and elsewhere.

In evaluating the popular impact extinction theory, several points need consideration. First, the Chicxulub structure is not known with certainty to be an actual crater. These large, unusual "cryptocrater" formations are numerous across the surface of the earth. Many of them may result from past geologic activity instead of impacts. Second, the observed iridium layer at the K-T boundary could be a dust deposit from volcanoes instead of from a space collision. Volcanic material generally includes trace amounts of the element iridium. In fact, a large region in western India called the Deccan Traps experienced massive volcanism at a similar relative time in history. This igneous event resulted in multiple layers of basaltic lava covering more than 500,000 square kilometers (180,000 square miles), to depths of a mile or more. Third, why did the alleged impact kill off the dinosaurs while many other forms of life remained healthy? Some of the least mobile creatures (tortoises, crocodiles) and also the most sensitive to climatic change (birds, fish) are still with us today. Fourth, the fossil record does not show an instantaneous demise of the dinosaurs. Some dinosaurs seemed to have died out within the Cretaceous period while others survived well into the Tertiary period, millions of years later in evolutionary thinking. Scientists have been unable to determine whether the K-T boundary layer itself was deposited in less than a year, or over tens of thousands of years. In view of these many problems, the impact view of dinosaur extinction remains uncertain, if not doubtful.

The possibility of a worldwide impact catastrophe is a popular topic today. For example, the collision of a Mars-sized object with the earth has been proposed to explain the moon's formation. Also, several dramatic movies have portrayed deadly asteroids or comets hitting the earth and endangering all life. In the evolution view there have been at least five major extinctions of life during earth history. These are called the Ordovician extinction (438 million years ago), Devonian (360 million), Permian (245 million), Triassic (208 million), and the "great dying" of the Cretaceous extinction (65 million years ago). The Permian event, as an example, is claimed to have eliminated 95 percent of all animal species living at that time. Some biologists believe we are now living during a sixth great extinction, this time brought on by man's domination of the earth. Creationists challenge the timescale that widely separates these extinction events. In the creation view there was one single major earth catastrophe of judgment that caused the near extinction of life, the Genesis flood of Noah's day. There also is a second, final judgment in the future when the heavens and earth will be cleansed by fire (2 Peter 3:10–12). Meanwhile, there is a definite future plan for mankind. We are not the result of chance, nor do we or the earth's animal and plant life face imminent extermination from a random asteroid collision.

---

6. How old is the earth?

Science alone is unable to pinpoint the exact time of earth's formation. Depending on interpretation, experimental data gives a wide range of age values ranging from very short to almost infinitely long. The meaning of such data is uncertain because of unknown rates of change and other variables in the past. Many creationists believe that the long-age results are flawed and that the physical evidence is strong for an earth and universe that are only 6,000–10,000 years old. Long extrapolations into the past generally are less reliable than the alternative data that supports a recent creation. One may wonder why there is such scientific zeal to project history billions of years into the past, while at the same time there is great hesitation to make scientific predictions just a hundred years, or less, into the future. The requirement for billions of years by the big bang theory and related evolutionary assumptions may be a leading factor.

In order to further refine the time of biblical origin, some have turned to the genealogies of Genesis 5 and 11. The years from Adam to Noah are listed, and also from Noah's son Shem to Abraham. As a fixed point, most scholars believe that Abraham lived around 2000 B.C. Men such as Johann Kepler, Bishop Ussher, and John Lightfoot long ago concluded from these Bible passages that the earth and universe were only about 6,000 years old. Other conservative scholars have urged caution in this "adding machine" approach to the Old Testament chronologies. It is uncertain whether the lists of patriarchal names are entirely complete or are meant to be representative. From this viewpoint the calculation of an exact creation date is not possible. Whether one accepts 6,000 or 10,000 years of history, however, both are many thousands of times shorter than the 4.6-billion-year figure commonly assumed for the earth's age. Several Scripture references have a direct bearing on the age question, as summarized in table 4.

Table 4

Selected Bible references that support a recent supernatural creation

Reference	Comment
Genesis 1, 2	The creation story is literal history. Chapters 1 and 2 are not in conflict, but complement each other.
Exodus 20:11	The normal days of creation are a clear pattern for our present calendar.
Matthew 19:4	Adam and Eve were present at the beginning, on the sixth day, not after eons of time.
Romans 5:12	There were not long ages of animal struggle, disease, and death before the curse of Genesis 3.
2 Peter 3:8	God is above time. What appears to us to take much time could happen quickly.
2 Peter 3:10–11	In the *last days* the entire universe will be transformed quickly, not over billions of years.

Some might ask why the earth looks old if it is actually quite young. In truth, appearance is of limited value in determining age. Whether for people or for the earth, looks can

be very misleading! Consider the volcanic island of Surtsey that formed off the coast of Iceland in 1963. Initially
the newly formed land was covered with hot basalt, supported no life, and had no appearance of age. Within days,
however, wave action eroded the freshly hardened lava into
an ancient-looking coastline with a wide beach. Birds
arrived within a year, bringing with them seeds and insects.
Larger animals and other vegetation soon were carried to
the island by driftwood. Just four decades after its birth,
Surtsey now has the appearance of a semimature island. A
visitor to the island might guess wrongly that it is many
centuries old.

Many regions of the earth show similar evidence of dramatic, sudden change instead of slow modification over
inactive eons of time. As a further example, consider the
channeled scablands of eastern Washington state. In this
region there are 16,000 square miles of cliffs and canyons.
For many years this landscape was considered to have
formed very slowly as gradual erosion took place. Now geologists know that the actual story is far different. At the end
of the ice age just a few thousand years ago, great inland
lakes existed across the United States. In particular, Lake
Missoula stretched across eastern Washington, Idaho, and
western Montana. When a giant *ice dam* broke, the entire
lake (the size of Lake Michigan) rushed across the land to
the Pacific Ocean. This local flood moved at 100 miles per
hour and eroded channels 200 feet deep in solid basalt.
Boulders as large as houses were tumbled about by the water,
and streambed ripple patterns of soil 20 feet high were
formed and still exist today. It was all over in just a few days,
leaving the channeled scablands as a landscape memory of
the dramatic event.

Even the Grand Canyon, often called a "monument to
time," displays startling youthfulness in the Genesis flood
interpretation. An important factor in measuring time peri-

ods is the *rate of change*. If geologic changes must occur very slowly, then there is indeed evidence of long age. If past changes have happened quickly, however, whether by flood, erosion, or some other catastrophe, then the earth's age may be quite young. Large-scale erosion may require much time with *little* water, but only a little time is required with *much* water. Hundreds of geologists today, both secular and creationist, believe that rapid, catastrophic changes have shaped the earth's surface. On this basis a sound scientific argument can be made that the earth is around 10,000 years old, not billions of years old.

---

7. What happened to dinosaurs in the creation view?

Dinosaurs were made during the creation week along with all the other plants and animals. The land animals appeared on day number six. In preflood times, dinosaurs lived on the earth together with mankind. When the great flood came in the days of Noah, dinosaur representatives were present on board the Ark, along with all the other animals. Dinosaurs did not become extinct until sometime after the flood.

Climate changes occurred in the postflood world. For a period of time the earth remained warm, largely due to the vast ocean waters. Much of this water had come from underground when the "springs of the great deep burst forth" (Gen. 7:11). This subterranean water was warm in nature. In the early postflood years, animal populations again increased on the earth. During the following centuries, however, there developed a worldwide cooling trend. The loss of vegetation during the flood had resulted in much barren land. This sandy, rocky surface reflected sunlight upward rather than capturing it. An ice age slowly developed as average global temperatures dropped by 5–10 degrees Fahrenheit (2.8–5.6 degrees Celsius). Animals that could not adjust to this cooler climate, such as the tropical dinosaurs, gradually diminished in number. In the creation view, humans may also have hunted the last surviving dinosaurs. Historically, mankind has always hunted large animals. The biblical patriarch Job probably lived during this cooler postflood period of time. Interestingly, there are indications of a cool climate in Job 37:6–10 and 38:22, 29. These verses refer to snow, cold, ice, and hail.

When the earth's vegetation gradually returned, especially across the great equatorial rain forests, the earth's average temperature again increased. The continental ice sheets then receded toward the polar regions. However, it was already too late for many tropical animal species, which had succumbed to the colder climate. Note that it was not the Creator who was responsible for this loss of life. The flood itself was triggered by people who had turned their backs on their Creator.

One consequence of this creation scenario involves the earth's sedimentary rock layers. Instead of forming gradually, most were laid down quickly during the yearlong flood event. Thus some dinosaur fossils might be expected to be found in rock layers beyond the Mesozoic era, either earlier or later, where paleontologists would not expect them. On the few occasions when this has actually happened, the usual explanation is that the fossil strata has been somehow "reworked." That is, fossils were eroded from their original rocks and then redeposited in other rock layers.

Some may wonder about the multiple rock layers observed in locations such as Arizona's Grand Canyon. The geological complexity of the Genesis flood cannot be overstated. There has been no comparable earth catastrophe either before or since. During the yearlong flood, the moon was present in the sky. As earth's nearest neighbor, the moon is responsible for our tides. With a worldwide, shoreless ocean, the tides would have been quite dramatic. During the early flood stage, much of the upper surface material of the earth was drawn into water suspension. Then, in the latter flood months, this material began to settle out. The moon causes two high and two low tides every 24 hours at a given earth location. Thus one can imagine sedimentary rock formation being controlled by the tides depositing the material in multiple layers.

A prediction of the creation view is that there will not be another major ice age. If the colder climate was indeed triggered in the aftermath of the Genesis flood, then it was a singular event. We have God's rainbow promise of no more worldwide floods (Gen. 9:12–17). In contrast, geologists recognize at least four separate ice age episodes over a vast timescale. Creationists challenge these multiple events. The usual evidence, debris deposits, may have been transported by water instead of ice.

---

8. What are dinosaur "graveyards"?

Our modern cemeteries are places of order with their distinct rows and headstones. In fossil burials of dinosaurs, fish, and other animals, however, chaos often reigns. Bones from various animals are often packed together within rock, sometimes with the bones broken and distorted. In some fossil assemblages it appears that entire herds rapidly perished together.

A large dinosaur graveyard was found in 1907 in present-day Tanzania, Africa. During the next five years, 235 tons of this fossil freight, bones embedded in plaster and rock, were shipped to Berlin for preparation and display. Dinosaur National Monument on the Colorado-Utah border is the site of another large dinosaur bone bed. And at the nearby Cleveland-Lloyd quarry in eastern Utah, 10,000 dinosaur bones have been mined. All are scattered and out-of-place from their original skeletal positions. Dinosaur Provincial Park in Alberta, Canada, boasts another large dinosaur graveyard.

The usual explanation of these graveyards is that sudden, local flash floods inundated large herds of grazing dinosaurs. Other mass burial sites are said to result from animals attempting to cross flooded rivers. Such local disasters have undoubtedly happened in earth history. However, one should not overlook the greatest catastrophe of all time, the Genesis flood. During the yearlong flood, billions of drowned animals were moved about by turbulent water and then buried in flood deposits. This worldwide event helps explain the chaotic jumble of bones found in many dinosaur bone beds.

---

9. Did dinosaurs evolve into birds?

Paleontologist Martin Lockley writes with tongue-in-cheek that *Tyrannosaurus rex* is the great-grandfather of the hummingbird (Lockley, 1991). This dinosaur-bird connection was first popularized by evolutionist Thomas Henry Huxley as early as 1870. Supporters point to the theropods, small meat-eating dinosaurs. Most of these animals were 10 feet or less in length, moved about on two legs, and had small forearms. The theropods had a superficial ostrich-type skeletal appearance, which led to the idea that these dinosaurs eventually changed into birds. However, there are several serious objections to this idea.

First, there is no convincing evidence that dinosaur skin or scales ever mutated into feathers. Several birdlike fossils have indeed been found in China. And some appear to have feathers, such as *Caudipteryx*, *Protoarchaeopteryx*, and *Confucuisornes*. Others like the turkey-sized *Sinosauropteryx* simply had frayed, bristlelike collagen fibers under the skin, but no feathers. No example of scales changing into feathers ("sceathers"?) has ever been found—a convincing fossil link between dinosaurs and birds is still missing. And if some dinosaurs actually did have feathers, they may have been used for insulation rather than for flight.

Second, the "finger digits" of dinosaur forearms and corresponding bird wing bones show basic differences rather than similarity. Theropod dinosaurs also had heavy balancing tails and short forearms, an opposite structure from modern birds.

Third, the lung structure of the theropod dinosaurs resembles the bellowslike lungs of crocodiles instead of the high-performance lungs of modern birds. Birds' lungs are highly complex and unrelated to any other living animal.

Fourth, the eggs of dinosaurs and birds show marked differences. For example, bird eggs have a tough membrane within the shell. This membrane is lacking in many reptile eggs.

Fifth, the fossil bird *Archaeopteryx* ("ancient wing") causes a fundamental time conflict for dinosaur-bird evolution. First discovered in 1861, seven of these fossils the size of a blue jay are known to exist. *Archaeopteryx* has many features of modern birds including flight feathers and a wishbone. However, it supposedly lived during the early Jurassic period, 30–80 million years before primitive birds are thought to have evolved from dinosaurs. Another bird fossil found in the desert of west Texas in 1983, *Protoavis,* is dated even earlier, 75 million years before *Archaeopteryx.* How can these modern-looking birds possibly have existed long before birds themselves evolved from the dinosaurs? Obviously, there is *no* evolutionary consensus on the origin of birds.

The usual explanation for flight involves dinosaurs running and flapping their emerging wings. Then, one day, they lifted off the ground and flew. An alternative is that small dinosaurs climbed trees, then jumped and glided to the ground. Neither view has supporting evidence.

A chief evolutionary reason for believing that dinosaurs somehow became birds is that there are simply few other options for bird origin, apart from supernatural creation. In the creation view, dinosaurs and birds were distinct creatures from the beginning of time, true of all the distinct animal and plant kinds. Any apparent similarities between dinosaurs and birds show the common Designer, not a common ancestry (Monastersky, 1997). Birds were created on the same day (day five) as the flying reptiles and marine reptiles, and just one day before the land animals, including the dinosaurs.

Part Two
Dinosaurs and Modern Times

10. When did modern studies of dinosaurs begin?

In the creation view, dinosaurs were known to mankind in antiquity. Then in postflood times this knowledge was gradually forgotten. When large fossil bones were eventually found much later, they were often misinterpreted. For example, in 1677 a large bone was found in England. It was initially attributed to the giant humans described in Genesis 6:4. However, surviving drawings of this bone look similar to a dinosaur femur. Dinosaur fossil evidence was also seen during the western American expedition of Meriwether Lewis and William Clark. Clark's journal during 1806 notes a 3-foot rib found near present-day Billings, Montana. The explorers thought it belonged to a large fish, but instead it was probably from a stegosaur (Farlow, 1997).

In 1811–12, a series of severe earthquakes rocked midwest America. They were centered at New Madrid, Missouri, an area still subject to frequent tremors. Pioneer diaries from this early time describe the severe seismic events, including the following drama:

> People heard the rumble of thunder deep in the earth and the crash of chimneys. Fissures opened beneath people's feet and from yawning holes spouted jets of warm water, sand, and coal that flew as high as the treetops. In the morning we found, on the ground, the massive bones of giant creatures that had been long buried (Dohan, 1981).

It is possible that some of these bones ejected from deep underground belonged to dinosaurs.

In 1822, physician Gideon Mantell and his wife, Mary Ann, found large dinosaur teeth and bones while fossil hunting near Sussex, England. Dr. Mantell gave the name *Iguanodon* ("iguana tooth") to this new creature. William Buckland is credited with publishing the first scientific description of a dinosaur in 1824, the *Megalosaurus*. A member of another fossil-hunting family, Mary Anning, found the first British pterosaur (flying reptile) in 1828, now called *Dimorphodon*. Many additional large bones were found in the following years, attracting the interest of anatomist Richard Owen. He coined the term dinosaur in 1841 from the Greek roots *deinos* (terrible, fearfully great) and *sauros* (reptile, lizard). This name was chosen by Owen to engender awe and respect for these impressive creatures. Owen's "Report on British Fossil Reptiles" was presented at a conference during which he spoke for two-and-a-half hours! In this talk Owen declared that dinosaurs had not evolved but instead were distinct animals directly created by God. Concerning limited variation within animal kinds, Owen stated that the "Divine mind which planned the archetype also foreknew all its modifications." Ever since these more recent dinosaur fossil finds, the magnificent creatures have remained a popular area of scientific study.

The years 1890–1930 were a "golden age" for dinosaur exploration in North America. Many new fossils and footprints were found and placed in natural history museums around the world. A second golden age is occurring at the present time. There are many current discoveries and surprises being uncovered in the dinosaur fossil record all around the world.

---

11. Are there creationist paleontologists?

Yes, there are scientists in every research discipline who support the creation view, including fossil specialists. As an example, consider the early years of dinosaur discovery. Many pioneer researchers promoted the creation position in their work. William Buckland (1784–1856) was a British scientist who described fossil creatures two centuries ago. Around 1815 he found huge teeth near Stonesfield, England. As mentioned in the previous question, he later named this animal *Megalosaurus*. Buckland used paleontology to support biblical teachings. He taught that the fossils showed creative design, and that fossils resulted from burial in the Genesis flood (Farlow, 1997). Many creationists today agree with Buckland's conclusion.

Friedrich von Huene (1875–1969) was a German paleontologist who produced hundreds of publications. He did pioneer work on dinosaur fossils in Patagonia, southern Argentina, one of the most active research areas today. In 1887, von Huene was a major proponent of the division of the Dinosauria into Saurischia and Ornithischia branches, the "lizard-hipped" and "bird-hipped" classification widely used today. The son of a Lutheran pastor,

41

von Huene was deeply religious. He repeatedly declared that his research showed the intricacies of divine creation to those with eyes to see the truth (Farlow, 1997). Von Huene probably had in mind Romans 1:20, where creation evidence is said to be clearly visible in nature. The verse further states that people have no excuse for ignoring this evidence.

I will not attempt to name current creationist paleontologists for two reasons. First, their number is inevitably small since the evolutionary view has strictly controlled paleontological training in universities for over a century. Students with an interest in creation are normally screened out of the process. However, there are a host of amateur creationist paleontologists, some well qualified and experienced in this field of study. A second reason for not listing current names is confidentiality. Current professionals who prefer the creation view are often in sensitive career positions. In academia and in professional societies, job security may be at stake. Unfortunately, there is little tolerance for the creation view in such circles today.

---

12. Describe Dinosaur National Monument.

This site is located on the northern Utah-Colorado border. The monument consists of over 300 square miles of scenic wilderness and canyons drained by the Green River. Dinosaur bones were first discovered here by paleontologist Earl Douglass in 1909. Many of the fossils are imbedded in a sandstone rock layer that is tilted upward at 60 degrees. It is thought that the dinosaur fossils were once buried as much as a mile deep underground. Then they were later exposed by the upthrusting of the buried rock and subsequent erosion. Ten separate dinosaur species were eventually found in the formation. These include complete articulated fossils of *Apatosaurus* and *Diplodocus*. Articulated fossils are those in which the bones are found connected or in place just as they were in the living animal. Fourteen dinosaur skulls were found, which always are of great value in animal identification. The monument area is unique in its variety of dinosaur types.

The National Park Service left more than 1,600 fossil bones partially exposed in a patch of surface sandstone for educational purposes. These include both adult and juvenile dinosaurs. To protect the fossils from weather erosion and possible theft, a large glass enclosure nearly 300 feet long was built above a portion of the sandstone layer. Visitors may enter the enclosure to see and touch the embedded dinosaur fossils, now turned to stone. Dinosaur National Monument is an impressive fossil graveyard with some skeletal bones still in normal positions and others disarticulated. The site is traditionally thought to show rapid burial of the animals on a flooded river sandbar. The alignment of the disturbed bones even indicates the original direction of the water flow. It may be that burial actually occurred in surging waters during the initial stages of the Genesis flood.

The entire region for many miles around the national monument displays layers of sedimentary rock, some of which are hundreds or thousands of feet thick. Geologically, these rocks make up the Morrison Formation, said to be 150 million years old. This entire region covers 750,000 square miles of western scenery. It extends from Utah to Kansas and from Canada to New Mexico. At an earlier time the Morrison region was filled with wetlands, streams, and forests, a dinosaur paradise. At some later time, erosion cut through the Morrison rock layers, leading to the many beautiful canyons of the area. Other nearby rock formations include Arches and Canyonlands National Parks, and also the Flaming Gorge on the Green River.

13. Is there evidence that dinosaurs ever lived during human history?

There are indeed multiple evidences that people and dinosaurs once inhabited this world together—this in spite of denials by conventional science. None of these evidences provide proof with 100 percent certainty. However, the following partial list of items deserves serious attention.

1. There are many traditions involving monsters and dragons from the distant past. What is the origin of these stories? Critics of the creation view have suggested that ancient fossil bones and footprints gave rise to later dragon legends. However, it may be that pre- and postflood people actually interacted with such animals. As one example, there are worldwide stories of a unicorn-type beast. Though unicorns are commonly pictured as one-horned horses today, Chinese tradition describes the unicorn as a dragon with one horn. No modern land creature fits this single-horn description. However, a large herbivore dinosaur, *Monoclonius nasicorus,* had a single horn. Fossils indicate that this creature lived in both North America and Asia. The animal, related to *Triceratops,* thus may have been familiar to early people, and may have given rise to later unicorn traditions.

   Another example comes from the heroic poem *Beowulf.* This European epic was composed 13 centuries ago, and describes a still-earlier time in history. The story involves a dragon that ravages the land and is finally subdued by the hero, Beowulf. Parts of the story are certainly mythical. However, some of the characters, sites, and events in the poem can be historically verified. Perhaps the dragon was one of the last remaining dinosaurs. Scientist Carl Sagan once suggested that dragon stories were "fossil memories" that we still carry from our ancient ancestors millions of years ago. However, there is no known genetic mechanism by which memory can be inherited in this way. On the much shorter timescale of creation, dragon stories may have been passed down through the generations.

2. Petroglyphs are ancient drawings and paintings usually placed upon sheltered rock surfaces. They are found worldwide within caves and on cliff walls. Occasionally, pictures of dinosaur-like creatures are found. One such location is within Arizona's Grand Canyon. A petroglyph drawing of an apparent dinosaur was found by the Doheny scientific expedition in 1924. How did our ancestors know how to draw these animals unless they actually observed or remembered them?

3. In several locations worldwide there are what seem to be footprints of humans and dinosaurs in the same rock strata. If these are true human prints alongside those of dinosaurs, formed at the same place and time, the implications are momentous (see Question 44).

4. Detailed descriptions of dinosaur-type creatures are given in Job 40–41. The behemoth and leviathan creatures described in this Old Testament book clearly picture land and sea animals similar to dinosaurs (see Question 14).

5. A modern-day creature similar to some dinosaurs is the Komodo dragon. These monitor lizards live on several Pacific islands of Indonesia. Adult Komodo dragons are 10 feet long, weigh 300 pounds, have a poisonous bite, and carry large claws. They are not dinosaurs, but they surely have a similar appearance (see Question 27).

6. Over and over again, plants and animals thought to be long extinct have been found still living upon the earth (see Question 18). These *living fossils* show the fallibility of scientific pronouncements about extinctions in the distant past.

---

14. What were *behemoth* and *leviathan?*

These are Old Testament Hebrew names for specific animals. Their exact identity is somewhat obscure. The *behemoth* is named only in Job 40. Its description includes the following:

It feeds on grass like an ox (v. 15).
Its tail sways like a cedar tree (v. 17).
Its bones are like tubes of bronze and rods of iron (v. 18).
Behemoth ranks first among the works of God (v. 19).
The beast hides among the reeds in the marsh (v. 21).

Behemoth has been variously identified as a hippopotamus, elephant, or water ox. However, none of these animals fully fit the biblical description. Instead, the behemoth sounds very much like a large saurischian dinosaur, perhaps the *Brontosaurus (Apatosaurus)*. If true, then Job 40 provides the only detailed written description of this dinosaur. It should be noted that God Himself directly spoke the words of Job 40. Furthermore, Job was evidently familiar with behemoth. Since Job lived after the great flood, the *Brontosaurus* then did not disappear in the flood event. Instead, representatives left the Ark and began to repopulate the earth. A few behemoths still lived during Job's time. The name *behemoth* today has appropriately become associated with anything enormous in size or power.

Leviathan is mentioned several places in Scripture (Job 41; Ps. 74:14; 104:26; Isa. 27:1). In particular, Job 41 gives a full description. Here are some details about leviathan:

It cannot be captured (v. 1).
It is strong and fierce (vv. 8–10).
There are many teeth (v. 14).
Smoke and fire pour from its nostrils (vv. 18–21).
Its hide cannot be pierced (v. 26).
Leviathan's swimming leaves a wake (v. 32).

Modern identification attempts for leviathan often include a crocodile, whale, or some kind of sea serpent. Others believe that the leviathan must be a nonliteral, mythical animal that exists only in poetry. Contributing to this view is the reference to the fire-breathing nature of the leviathan. However, this behavior is entirely within the realm of possibility (see Question 48). The detailed biblical description of leviathan goes far beyond some imaginary beast; it closely matches that of the seagoing plesiosaur. This long-necked marine reptile glided through the water and certainly could leave a glistening wake behind. The people of Job's day were clearly familiar with both the behemoth and leviathan. Perhaps these impressive animals were some of the last remaining dinosaur and marine reptile examples living on the earth.

---

15. What did Japanese fishermen find in 1977?

A Japanese fishing vessel was operating in the southern Pacific Ocean east of New Zealand in April 1977 when an unusual carcass became entangled in the nets. The dead creature weighed 4,000 pounds, was 32 feet long, and displayed a long neck. After the crew took photographs and tissue samples, the animal's remains were returned to the sea. Ever since, there has been lively discussion about the creature's identity. Scientists have proposed that it was a mammal such as a whale, sea lion, or basking shark. However, the tissue analysis seems to rule out a mammalian origin. Also, the animal's anatomy does not clearly fit any kind of known shark. The animal's identity remains uncertain. A radical possibility is that the fishermen had snagged a recently expired plesiosaur. If this is true, then at least until 1977, a community of these marine reptiles still lived in the ocean depths. Over the centuries there have been many sightings of extraordinary sea creatures, some similar to the 1977 discovery. Modern marine reptiles would be the ultimate living fossil. Such evidence is a major challenge to evolutionary theories of time and extinction.

---

16. What is the Loch Ness monster?

Loch Ness is a deep, narrow freshwater lake near Inverness, Scotland. The lake measures 24 miles long, 2 miles wide, and is more than 1,000 feet deep in places. It is about 10 times smaller than Lake Erie, but 5 times deeper. Since the 1930s, when lake access was provided by a new highway, there have been many reports of a lake monster. Others have traced "Nessie" stories clear back to A.D. 600. The descriptions are similar to a 40-foot plesiosaur, one of the dinosaur-era marine reptiles.

The study of creatures commonly thought long extinct that may actually still be living is called *cryptozoology*. Examples of other "living fossils" include the coelacanth fish, the cattlelike okapi, and the tuatara reptile (see Question 18). In 1977, Japanese fishermen also hauled in a mysterious plesiosaur-type carcass near New Zealand (see Question 15).

There may indeed have been a recent, small population of marine reptiles living in Loch Ness. According to Genesis 1:20, all sea creatures were made on the fifth day of the creation week. It is only evolutionary speculation that assumes such animals became extinct millions of years in the past. Thus far all efforts to locate Nessie have failed. The many historical sightings and blurry photographs are sometimes attributed to large eels, seaweed, or mere hoaxes. There have undoubtedly been many of the latter.

In 1987, dozens of boats and a minisubmarine thoroughly searched Loch Ness with underwater cameras and sonar, once again without success. One major problem is the dark peat-stained water, which limits visibility to only a few feet. Another problem is the lake's maze of deep, unexplored underwater caverns. The River Ness also connects the lake with the North Atlantic Ocean. Thus there are abundant hiding places and possible escape routes for unknown lake creatures. Will Nessie ever be found? If it is real, the recent high-tech searches may unfortunately have driven this shy creature to the point of extinction. In any case, the history of Loch Ness sightings remains a serious challenge to the evolutionary timescale, which assumes no plesiosaurs have lived on the earth for 65 million years.

17. What is *Mokele-mbembe?*

Dinosaur sightings have been reported from Africa for centuries. Recurring stories come from the Likouala region of the Republic of Congo. This frontier area contains more than 55,000 square miles of remote jungle, swamps, and lakes. It is the size of the entire state of Georgia. Sparsely inhabited by pygmies, the region is largely unknown to outsiders. Unusual animal sightings center on Lake Tele, located 400 miles north of Brazzaville.

The Africans describe a large aquatic reptile that they call *Mokele-mbembe,* pronounced mo-Kay-lee em-BEM-bee. As one might expect, it is regarded as somewhat sacred and mythical by natives. Their drawings resemble a small *Apatosaurus* with smooth skin and a long tail. Numerous eyewitness reports extend from the 1940s through at least 1990. The beasts, if real, are rare and retiring.

Retired University of Chicago professor Dr. Roy Mackal conducted expeditions in search of African dinosaurs in 1980 and 1983. Mackal reported fresh footprints but no visual sightings. These searches were partially funded by the National Geographic Society, which then declined to publish the related story. Mackal explains that the Society wanted additional evidence beyond alleged footprints before publication. Mackal's experiences were recorded in his 1987 book *A Living Dinosaur?* Dinosaurs are usually thought to have become extinct 65 million years ago. Thus evolutionists do not appreciate Mackal's conclusion that "prehistoric" creatures still inhabit remote African regions (Mackal, 1987).

---

18. What "living fossils" are left from the dinosaur era?

A living fossil is an organism that still survives, although closely similar fossils are found in "ancient" stone. It is also defined as a thriving plant or animal that was once thought to be long extinct. The term was first used by Charles Darwin in 1859 in his *Origin of Species.* As new living plants and animals are discovered in remote locations, the number of living fossils continues to increase. Each example is a glaring exception to evolutionary change. Several living fossil descriptions follow.

Ants have been found entombed in amber from New Jersey and dated at 92 million years old. This vast age is rejected in the creation view, but evidence shows that ants indeed lived during the dinosaur era. The fossils show very little change from modern examples of ants (Agosti, 1997). Ants have always played an essential part in the health of ecosystems. Today in the Amazon rain forest, ants make up more than 25 percent of the total animal biomass. According to Proverbs 6:6, ants are valuable as a practical example of ambition.

The *chambered nautilus* is a deep-sea mollusk with a soft body enclosed in a spiral shell. Like their squid cousins, these nautiloids move by taking in water and then forcing it outward as a jet. Living in the western Pacific Ocean, nautiloids swim upward from great depths at night to feed on surface microorganisms. Fossils of the chambered nautilus and other nautiloids are dated at over 100 million years old. Hence the secretive and beautiful nautilus qualifies as a living fossil. Some evolutionists have suggested that nautiloids must evolve or change over time, but that they somehow keep returning to the same "primitive" form.

The *coelacanth* is a fish whose fossils are found from Devonian period strata up through the Cretaceous period (see table 1). The name means "hollow spine," describing the supports of the creature's dorsal fins. Coelacanths were thought to have already existed 200 million years before dinosaurs appeared, then died out about 80 million years ago. However, in 1938 fishermen caught a live coelacanth in the Indian Ocean near the Comoro Islands north of Madagascar. The coelacanths living today look exactly like their fossil counterparts. More recently, additional coelacanths also have been found living in Indonesian waters. Therefore, this creature may be more common around the world than previously thought. This impressive fish grows to a size of 6 feet and can weigh 150 pounds. Not at all primitive, the coelacanth bears live young instead of eggs, and boasts complex electroreceptor abilities for the detection of prey. There is no evolutionary trace of coelacanth fossils from the supposed intervening millions of years between the "ancient" fossils and the living specimens.

Consider two additional comments about the coelacanth. First, the Comoros, their habitat region, are recent volcanic islands existing for "only five million years" or less. How could coelacanths have continuously lived in this region of turbulent, evolving crust for the assumed 65 million years since the end of the Cretaceous period? There is a basic time problem here. Second, the coelacanth is a complex live-bearer rather than an egg layer. When non-egg embryos were found inside a female fossil coelacanth in 1926, experts concluded that the embryos must have been *eaten* by the mother. After all, they reasoned, such "primitive" fish must lay eggs. We now know better.

Ginkgo leaves were fossilized in the Jurassic period at the time of the dinosaurs, supposedly 208–65 million years ago. These fossilized leaves are found widely across Australia, North America, and Europe. However, for millennia the living ginkgo tree, indistinguishable from its fossils, has grown in Asia. In modern times, ginkgo trees have been planted worldwide in cities and parks because of their resistance to urban pollutants. Extracts from ginkgo leaves have been found useful in the treatment of hearing loss, poor circulation, and Alzheimer's disease. In fact, nearly half the medicines known today were similarly discovered as chemicals in microorganisms, plants, or animals. This is surely a bountiful heritage from the creation. Along with the living fossil ginkgo, sequoia and cycad trees also have changed little over "geologic" time.

Horseshoe crabs have a large, rounded body and a stiff, pointed tail. They are abundant in the seas today, and identical specimens are found in the fossil record. Horseshoe crabs are supposedly dated to be at least 200 million years old, from the Jurassic period. With a highly complex circulation system, these arthropods are not simple or ancestral, as sometimes stated. A valuable reactant extracted from their blood is commonly used to test drugs and medical instruments for contamination. Healthy horseshoe crabs are kept in laboratories to be used for this purpose.

Congo tribesmen long described a large, hoofed animal with a giraffelike head and zebra stripes. Paleontologists found fossils fitting this description, but they assumed that the mammal had been extinct for 30 million years. However, in 1906 British scientist Harry Johnston captured the creature, called an *okapi*, alive. These big cattle are now on display in many large zoos. The okapi shows that even large animals have escaped detection until modern times.

The *triops* is a species of tadpole shrimp. About 1 inch long, they are thought to have been a food source for dinosaurs since triops fossils encompass the "Age of Reptiles." Today living triops can be found in ponds in the southwest United States. One can also purchase a "triops kit" from educational suppliers and raise these small creatures at home. Their dried eggs, when moistened, hatch to produce living fossil shrimp.

The *Tuatara* is a lizardlike reptile that lives on rocky islands off mainland New Zealand. Also called a *sphenodon*, the adults are 2 feet long, and they often live more than 100 years. Their fossils are found in abundance with dinosaur bones. However, there are no known *Tuatara* fossils dated younger than 80 million years. The evolutionary predicament is that this still-living creature somehow completely disappeared from the fossil record for 800,000 generations.

The large and growing list of living fossils also includes the following organisms:

Dragonflies Their smaller descendants still live today. These slender insects have two pairs of wings. Large-size specimens with

2-foot wingspans were common during the Carboniferous geologic period, said to be 100 million years before the dinosaurs arose. Dragonflies are by no means simple or primitive. Their digestive tract, for example, is a perfectly functioning miniature chemical plant.

Sea lilies or crinoids These have recently been discovered living in deep-sea trenches where the ocean water is several miles deep. Previous to this finding, crinoids were known only as "ancient" fossils.

Stromatolites These are pillarlike sedimentary structures that are coated with microorganisms. The surface bacteria trap sand grains in shallow sea water to build pedestals several feet in size. Stromatolites are thought to have been the most abundant form of life during the Precambrian era.

Thermophiles These aquatic microbes are found in Yellowstone National Park in Wyoming. They thrive in water that is hot enough to burn a person. Like the stromatolites, these hardy bacteria are thought to date back to the earliest life in the Precambrian era.

Wollemi pine trees Living in Australia, these trees have no known fossils younger than a supposed 65 million years. They may be seen in Wollemi National Park near Sydney.

The list of still-living organisms from the dinosaur era goes on indefinitely. Here are additional examples, some with the number of years they are thought to have lived unchanged on earth.

bacteria (3.5 billion years)
blue-green algae (3.5 billion years)
cockroach (250 million years)
coral *Heliopora* (100 million years)
crocodile (140 million years)
cycad plant (250 million years)

duckbill platypus (150 million years)
feather mites (120 million years)
frog (275 million years)
gar (50 million years)
horsetail (300 million years)

lamprey
lancelets (550 million years)
lingula brachiopod (600 million years)
lungfish (200 million years)
magnolia tree
metasequoia tree (40 million years)
monkey puzzle tree
mouse deer *Tragulid* (50 million years)
mussel
neopilina mollusk (500 million years)
opossum (70 million years)
oyster

parrot (70 million years)
peripatus worm (500 million years)
ray
scallop
sea urchin
shark (400 million years)
shrew
silverfish (350 million years)
spider (400 million years)
tapir (38 million years)
tortoise, turtle (275 million years)
tree squirrel *Sciurus* (35 million years)
welwitschia plant

In the creation view, all of the created kinds of plants and animals lived together from the beginning of time. Therefore, there is little surprise at the growing list of discoveries of modern life-forms that also lived with the dinosaurs. Each living fossil is a testimony to the creation, and each is a dramatic exception to the theory of evolutionary change.

Part Three
The Dinosaur Family

19. What is a dinosaur?

A dinosaur definition is not as simple as it might seem. Some simply identify dinosaurs as extinct reptiles, which were often large in size. If dinosaurs were warm-blooded, however, then they were not true reptiles as usually defined (see Question 38). Others define dinosaurs as lizardlike creatures from the distant past. However, the upright posture of many dinosaurs appears to have been distinctly unlike modern lizards, which have a sprawling gait.

Paleontologists today define dinosaurs with a whole *suite* of technical characteristics. These include multiple sacral vertebrae, a strongly asymmetric manus (forelimb bone), and a slightly S-shaped third metatarsal (foot bone). The families of marine reptiles and flying reptiles are not considered dinosaurs, although they lived on earth at the same time. Perhaps the safest dinosaur definition is that these majestic creatures made up several distinct dinosaur *kinds*, the generic word used in Genesis 1 for separate animal and plant categories. The biblical *kind* probably is similar to the biological classification of genus, family, or subfamily. Thus there were created, as examples, the dog, cat, and dinosaur kinds.

_____

20. How are dinosaurs classified?

In brief, dinosaurs can be thought of in three major categories. Theropods are the two-footed, three-toed meat eaters such as the *Tyrannosaurus*. The sauropods are the giant elephant-like beasts that moved about on four legs, like the *Brontosaurus*. They were the largest terrestrial animals ever to walk the earth. Both theropods and sauropods are called saurischian dinosaurs. Finally, the ornithopod dinosaurs have a pelvic structure similar to birds. This group includes the duck-billed dinosaurs, and also the armored and horned dinosaurs. Keep these three categories in mind and your dinosaur studies will be simplified. The following paragraphs give additional details on dinosaur varieties.

Dinosaur classification continues to be an unresolved topic among experts. There are almost as many versions as there are paleontologists. Many modern animals are identified and categorized on the basis of their skin or feathers. However, dinosaur fossils are largely limited to bones, often giving an uncertain identification. Dinosaurs are placed in the biology classification scheme shown in table 5, based on their assumed appearance. The general class of reptiles includes snakes, lizards, crocodiles, and turtles. Reptiles are identified as cold-blooded, usually egg-laying vertebrates with a covering of scales or horny plates. There are many reptile subclasses including the archosauria ("ruling lizards"). General defining features of the archosaurs include teeth set in individual sockets and a skull with two openings near each temple.

Of the five archosauria orders, two are properly called dinosaurs, the saurischia and ornithischia. The main difference concerns the bone structure of their hipbones and pelvis. Saurischia ("lizard-hips") include first, the carnivorous theropods ("beast-footed") like the *Tyrannosaurus.* They walked on two legs, each with three birdlike toes. Most theropods also have bladelike teeth with serrated edges. Second are the sauropods ("lizard-footed"), often herbivorous and gigantic in size, such as the *Brontosaurus.* Sauropods have small skulls and long necks and tails. They are quadrupedal, traveling on four massive limbs, each with five toes.

The second major order of dinosaur is the ornithischia or bird-hipped variety. Their pelvic bone has an extra prong when compared to the saurischia (see figure 3). The ornithischian dinosaurs were typically herbivores. Many of them were small, bipedal, and more diverse than the saurischia. However, exceptions are often the rule in dinosaur definitions. Some of the ornithischia members were quadrupedal and large. The bird-hipped ornithischia include four suborders (table 5). These are the ornithopods ("bird-feet") such as *Iguanodon* with no armor, the armored ankylosaurs, which were covered with bony plates, stegosaurs with large raised triangular plates along their backs, and the ceratopsians with horns like *Triceratops.*

Dinosaur research and discovery is a very active area of paleontology. Ten or more new dinosaur species are being discovered annually around the world. The total number of distinct dinosaur species is now about 1,000. God evidently created the dinosaurs in great abundance and variety.

Species names are written as *Genus species,* following the taxonomy of Carl Linnaeus (1707–1778). His hierarchy of life divisions include kingdom, phylum, class, order, family, genus, and species. The dinosaur species name is abbreviated as *G. species,* as in *T. rex* for *Tyrannosaurus rex.* Most dinosaurs are commonly known only by their genus name.

Figure 3

(a)

(b)

Comparison of the typical pelvic bones of saurischian ("lizard-hipped") (a) and ornithischian ("bird-hipped") (b) dinosaurs.

Table 5

A partial classification scheme
for reptiles including dinosaurs

Class	Subclass	Order	Suborder
reptiles	archosauria		
		saurischia	theropods
			sauropods
		ornithischia	
			ornithopods
			ankylosaurs
			stegosaurs
			ceratopsia
		thecodonts	
		crocodiles	
		pterosaurs	

The dinosaur part of the list is shaded.

21. Is there an evolutionary ancestor of dinosaurs?

It is commonly thought that all the dinosaurs descended from small, bipedal, meat-eating animals. These ancestors are variously called pseudosuchins, paleotetrapods, or thecodontia (Padian, 1986). This latter name means "socket-teeth." One particular fossil that is promoted as a dinosaur precursor is *Eoraptor* ("dawn stealer"). It is placed in the late Triassic period, 230 million years ago, prior to the time of dinosaur dominance. *Eoraptor* was a swift animal of about

25 pounds, about the size of a turkey. Its fossils show none of the horns or armor typical of many dinosaurs. It is assumed that the evolutionary mechanisms of mutation and natural selection gradually gave rise to the various dinosaurs with their distinctive features. Mutations are permanent random changes in the DNA within cells. These mutations, actually mistakes, may result from exposure of the parent to radiation or chemicals. Since these mutations occur in reproductive cells, changes will be passed on to future generations. Natural selection then follows as a "weeding out" of unhealthy mutations and the passing on of beneficial mutations to future generations.

There remain at least three major missing links in this assumed dinosaur lineage. First, where did the *Eoraptor* ancestor itself come from? Natural origin theories, whether for dinosaurs, for life in general, or even for the moon, always have an unexplained starting point. Second, where are the many expected fossil connections between *Eoraptor* and the dinosaurs? These multiple links are still missing. It is likewise assumed that early reptiles gradually evolved into modern mammals, but once again the fossil links cannot be found. Third, experience has shown that mutations are harmful and debilitating. Beneficial mutations are at best rare, and perhaps nonexistent. Rather than leading to evolutionary progress, natural selection is instead a conserving process at best. It screens out harmful genetic problems to protect the health and stability of the created kinds.

---

22. Describe some of the popular dinosaurs.

Several of the better-known dinosaurs will be described here in alphabetical order. *Allosaurus* weighed up to 4 tons and must have been a formidable predator. Its teeth had serrated edges like steak knives. It reached 40 feet (12 meters) in length and stood 15 feet (4.5 meters) tall. *Allosaurus* walked on hind legs and used its massive tail for balance. Fossil remains have been found in Colorado and Wyoming.

Ankylosaurus means "crooked lizard," alluding to the large curved ribs of this dinosaur group. The ankylosaur reached a length of 25 feet. It was a short-limbed, heavily armored animal somewhat like a modern armadillo. Like a "living battletank," its entire back was enclosed in bony plates and long spikes. At the end of its tail was often a bony knob, perhaps used as a defensive club. Vascular grooves covered the surface of the ankylosaur armor. These indicate many capillaries and a rich blood supply. The armor may have had a pink tint as a result of this circulation of blood within the bony covering. Thus the ankylosaur may have "blushed" (Farlow, 1997).

Apatosaurus or *Brontosaurus* was a long-necked giant of its day. This creature reached a length of 80 feet and weighed 35–45 tons, half the weight of a large whale. Fossil locations imply that the apatosaurs and other large sauropods preferred dry terrain to swamps, the reverse of earlier thinking. Older drawings often showed these large dinosaurs almost completely submerged with only their long necks held above water. However, if this were the case, water pressure would probably have collapsed their lungs. The head of *Apatosaurus* was somewhat similar to a horse's head in size and shape. Nerve and blood pathways in the skull suggest that the *Apatosaurus* had an extremely sensitive snout, like modern animals.

Diplodocus was one of the large, somewhat barrel-shaped sauropods. As an adult it displayed a long neck and tail, about 23 feet (7 meters) and 46 feet (14 meters) long respectively. Like the apatosaur, *Diplodocus* was once thought to live mainly in water, walking on the lake or river bottom with its neck extended above the surface. However, this would place great pressure on its lungs, positioned more than 20 feet (6 meters) beneath the water surface. More likely it was a land animal, with a strong heart that pumped blood to twice the head height of modern giraffes. There is little fossil evidence that any of the land dinosaurs spent much time in water.

Hadrosaurus simply means "big reptile." This name includes the general family of duck-billed dinosaurs. They had a flattened, toothless duckbill with multiple rows of small teeth behind. *Hadrosaurus* could chew food somewhat like a carrot grater. A duckbill fossil found in Wyoming also revealed evidence of webbing between its toes and a wide tail like a crocodile. Many hadrosaurs had crests and bumps on their heads. Some skin impressions in rock show a pebbled appearance similar to that of an elephant's skin. Other impressions alternately suggest a scaly skin like a tortoise or lizard. The stomach contents of another duckbill find revealed that they ate food on land. That fossil's last meal consisted of conifer needles, twigs, and seeds from land plants. Reaching 40 feet in length, it appears from the muscle structure that duckbills may have occasionally ventured into the water as powerful swimmers.

The duckbill dinosaur named *Parasaurolophus* displayed a large bony head crest. This protrusion has variously been identified as a snorkel, air storage chamber, butting organ, resonating chamber, and a scent detector. Quite apparently, no one knows the true reason for this hump! If it was a horn, fearsome sounds may have been this hadrosaur's main defense mechanism. These dinosaurs could boast no armor plates, long claws, spikes, or whiplike tails.

In 1998, a hadrosaur fossil was found in the southwest United States and dated at 100 million years old. This find has bewildered paleontologists, who have long believed that hadrosaurs originated in Asia, then first migrated to North America less than 90 million years ago.

Iguanodon fossils were first found in 1822 in England, beginning the modern era of dinosaur study. Their tooth structure implies a plant diet. *Iguanodon* tracks show that they traveled in groups, probably for safety as they grazed together. Adults reached a length of about 25 feet. A fossil spike or spur several inches long was initially thought to be positioned on the *Iguanodon's* nose, similar to a rhinoceros. More complete fossils have shown that the spikes are actually located at the thumb position instead of the nose. These stout daggers would have been formidable weapons of defense.

Pachycephalosaurus was the dinosaur equivalent of the ram or bighorn sheep. This rare dinosaur had a massive skull with solid bone material as much as 9 inches thick. It may have butted heads with its friends in competition for dominance. They would probably have exchanged glancing side blows to prevent neck injury. Others suggest that the thick domes were used to butt the flanks of opponents.

Spinosaurs were a group of long-snouted, narrow-mouthed dinosaurs. These theropods had a head somewhat like a crocodile. One example is named *Suchomimus,* or "crocodile mimic." It may have hunted fish or other small dinosaurs. Many *Spinosaurs* also had impressive sail-like fins on their backs, taller than a person.

Stegosaurus is best known for the large, triangular bony plates along its back and tail, ending with several pairs of tail spikes. Some of the largest back plates are 3–4 feet tall. They were not connected directly to the backbone, so their exact position is uncertain. The plates have been reconstructed in different patterns, either as a single row, two pairs of matched rows, or two alternate, staggered rows. There actually may have been different arrangements of plates in various stegosaur species (see figure 4). These back plates helped the animal cool its body (see Question 46). In addition, these extensions may have played a part in courtship display. The plates also would have made *Stegosaurus* look larger, which is useful as a defensive mechanism. It also has been suggested that this animal may have resisted attack by curling up like a giant hedgehog with its plates pointing outward and its tail lashing sideways. The length of *Stegosaurus* reached 20 feet, and the adults weighed about 5 tons. Muscle structure and center of gravity studies show that *Stegosaurus* was capable of rearing upward on its hind legs in a cranelike posture to reach treetop food, as elephants sometimes do.

Figure 4

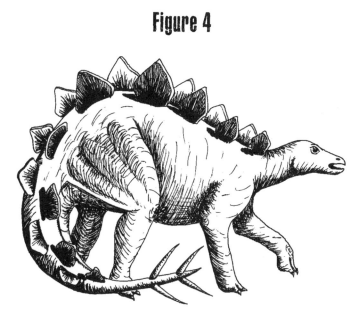

Stegosaurus, showing one possible orientation of its back plates, two alternating rows.

Triceratops, or "three-horned face," had two front-pointing horns as long as 4 feet, and also a shorter nasal horn. Some of these horns show a graceful double curve similar to those of modern longhorn cattle. Whether the horns were limited to males or also grown by females is unknown. Puncture wounds are found in some skulls, indicating serious shoving matches between them. The creature also had a very large saddle-shaped neck shield made of bone. *Triceratops* weighed up to 10 tons, several times the weight of a modern rhinoceros. Its unusual jaw allowed the upper and lower teeth to slide vertically past each other like scissors. These jaws were operated by impressive muscles 3–4 feet long. Thus far, *Triceratops* fossils have been found only in western North America.

Tyrannosaurus, or "tyrant lizard," has a ferocious reputation. It is characterized by a large head, teeth as long as 12 inches, and the most powerful jaws of any known animal. The "*T. rex*" grew to a height of 20 feet and weighed 5–6 tons, similar to an elephant. As with other animals from the past, *T. rex's* actual diet is unknown. It may have been a fearsome predator, and is sometimes called a "land shark," which attacked prey at high speed. Alternatively, it may have largely scavenged meals from carrion. A complete *Tyrannosaurus* skeleton nicknamed "Sue" was auctioned off in 1996 to the Field Museum of Chicago for $8.4 million. It had been found in South Dakota some years earlier. Less than two dozen *T. rex* fossils have been discovered, few of them complete. They are thought to have lived during the late Cretaceous period, long after earlier dinosaurs such as stegosaurs were gone. One *T. rex* mystery concerns its small forearms, weaker than our own arms. These limbs appear too small to be useful to such a beast. In the creation view, however, everything in nature has a plan and purpose. The *Tyrannosaurus* forelimbs thus may have been useful for balance, or perhaps for getting up again after lying down to rest.

The fossil of *Gigantosaurus,* found in Argentina and Morocco, is similar to *T. Rex,* only larger. This creature grew to be 47 feet long and had a 6-foot skull. It may have been the world's largest carnivore. Older tyrannosaur drawings and museum mounts showed the tail on the ground as a supporting prop, with the head held high in the air. Many paleontologists now believe the bodies of tyrannosaurs and similar dinosaurs were more likely positioned horizontally. The tail then extended outward above the ground as a counterbalance to the body, with the legs acting as the pivot or balance point.

The first skull of a *Velociraptor* was found at the Flaming Cliffs location in Mongolia in 1923. This creature had formidable claws. It stood about 6 feet tall and had a narrow head. Another unusual *Velociraptor* fossil was found in Mongolia's Gobi Desert in 1972. It appears to have been in a death struggle with a *Protoceratops*. Both creatures are fossilized, with the *Velociraptor's* forelimb caught in the jaws of the large herbivore. It is not known whether the raptors were mainly scavengers or active hunters.

With about 1,000 known dinosaurs species, the above list is limited to very few examples. The incredible variety of these majestic creatures clearly displays God's creative imagination. In all likelihood, some of the most unique dinosaurs remain completely unknown today.

23. What are some problems of dinosaur reconstruction?

Dinosaur fossils are seldom found with bones in their natural positions. Instead, the bones are *disarticulated* and scattered randomly, sometimes miles apart. Fossils also are usually incomplete, and different dinosaur species commonly are mixed together. Most bones are carefully photographed and mapped before they are disturbed. Some fossilized bones are found in loose soil and can be removed easily. Other bones are embedded in solid sedimentary rock. This is the case at Dinosaur National Monument where the bones must be painstakingly removed from a hard sandstone matrix. A single dinosaur skull on display may represent many months of careful work removing it from its surrounding rock.

Several humorous mistakes have occurred in the assembling of dinosaur skeletons. For example, originally the wrong skull was placed on *Brontosaurus*, where it remained for almost a century (see Question 24). An even greater error was made with the marine reptile *Elasmosaurus*. In this case the skull initially was placed at the wrong end of the beast—at the tip of its tail. The tail had been mistaken for the neck! A large horn was once placed upon the snout of the *Iguanodon* fossil. It is now known to have been a thumb bone of the animal. The clawed toe of *Hypsilophodon* was initially placed in a backward, reversed position. It was thought that this dinosaur climbed trees, similar to certain tree-dwelling kangaroos. The claw is now known to point forward to provide traction while running.

Dinosaur bones often were strewn about by predators and scavengers before fossilization occurred. And after burial, overlying sediments sometimes distort or crush the bones. Some dinosaur bones also suffer from "pyrite disease." This is a mineral change within the bones that causes them to gradually disintegrate into powder. Iron pyrite is the mineral iron sulfide, FeS_2, which is often present in fossil bone and petrified wood. This yellow, metallic mineral is also known as "fool's gold." Pyrite disease is an occasional problem with mounted dinosaur skeletons in museums. Over just a few years the fossil bones may deteriorate into tiny fragments. Dinosaur bones also may be radioactive, a possible hazard for paleontologists. Uranium atoms from circulating water tend to accumulate in the fossils while they are underground. The bones may then remain radioactive for a long period of time.

Still today there are uncertainties in correctly assembling dinosaur bones in the laboratory. Large bands of fibrous tissue once existed between the bone joints. These ligaments were not fossilized and are missing. It is therefore not known exactly how these bones were originally oriented and connected together.

---

24. Why did *Brontosaurus* become *Apatosaurus?*

In 1877, some dinosaur vertebrae were found at Morrison, Colorado. Paleontologist Othniel C. Marsh of Yale University named the creature an *Apatosaurus,* meaning "deceptive reptile." Two years later many similar bones were found at Como Bluff, Wyoming. Marsh called this new dinosaur a *Brontosaurus,* or "thunder lizard." As additional bones accumulated in the following years, the two dinosaur fossils were realized to be very similar. According to the established "law of priority," the first name published with an adequate description becomes the permanent official name. The name *Apatosaurus* thus won over *Brontosaurus* by two years! Although a name correction to *Apatosaurus* was first made in 1903, the name *Brontosaurus* is still commonly used since it is one of the most familiar dinosaur titles. The two names are used interchangeably in many books, including this one.

There is a somewhat similar story for a small dinosaur named *Troodon* found in 1856. Nearly a century later, another new fossil was called *Stenonychosaurus.* It is now realized that both are the same species, properly called *Troodon.*

Early *Brontosaurus* fossils were found without skulls attached, a somewhat common occurrence. The first mounted *Brontosaurus* was given a skull that had been found several miles away. This particular skull had spoon-shaped teeth, and became generally accepted as correct for these dinosaurs. From the study of similar fossils, it was realized in about 1985 that the original *Brontosaurus* had a lighter, long-snouted skull with peglike teeth. *Brontosaurus* had been given the head of the wrong dinosaur, one that belonged to a *Camarasaurus!* Camarasaurs are the most common North American sauropod.

---

25. What plant forms existed at the time of the dinosaurs?

The Bible states that plants were created on the third day of creation (Gen. 1:11–13). This was two days before animals appeared and also one day before the sun was made. The creation week was obviously a time of supernatural events completely beyond our understanding. Plants initially thrived under some form of pre-sun light source. Light, after all, was created on the first day (Gen. 1:3). In the biblical view the dinosaurs, other animals, and mankind lived in the presence of all the created plants, including modern varieties.

Plant fossils indicate that the preflood dinosaur era was dominated by evergreens, which include pines and firs. Other common trees included short cycads with palmlike leaves, and cycadeoid and lepidodendron trees with diamond-shaped markings on their trunks. The ginkgo tree with its fan-shaped leaves was also present. Interestingly, living ginkgo trees today look exactly like their fossil ancestors (see Question 18). On the ground grew abundant ferns and horsetails, some reaching tree size. Maple and oak trees, thought not to exist with the dinosaurs, may have been present only in small numbers at this early time.

The evolutionary view allows few flowering plants, or angiosperms, during the dinosaur era, and no grasses. Instead, gymnosperms were supposedly dominant. These are plants with seeds unprotected by seed vessels. Much later, plants somehow evolved the ability to make flowers. These flowers then attracted insects and other pollinators. However, the assumed absence of flowering plants around 65 million years ago has been challenged in recent years. In 1990, the fossil of a flowering plant was found in Australia and was dated at 115 million years old. Then in 1998, a new fossil from China, related to the flowering magnolia tree, was dated at 140 million years old. For many people, the creation view with its fully functioning earth provides a refreshing alternative to the conflicting speculations of secular science.

The discussion of dinosaur-era plants raises an additional problem with evolution. Plant fossils are rarely found in the vicinity of dinosaur remains. This is true worldwide, from Montana to Mongolia's Gobi Desert. Even pollen and spores are found only in small numbers. Dinosaur diets would have required great amounts of vegetation. If dinosaurs were fossilized over millions of years, why not plants? This finding may indicate that the dinosaurs were transported from their original locations and deposited rapidly by water and mud during the Genesis flood.

---

26. Why aren't human or modern animal fossils found with dinosaur fossils?

Dinosaur bone beds are indeed lacking in mammal fossils. No large mammals are found, for example, at Dinosaur National Monument. To understand why, one must keep in mind the conditions of fossilization. When groups of dinosaurs died and were buried, perhaps in the initial stages of the great flood, it is doubtful that creatures such as bears or lions would have been in their midst. That is, the observed distinct fossil layers may well show ecological sorting based on animal locations. This explanation contrasts with the evolutionary view that mammals arrived much later in history than dinosaurs.

As a further example, consider the La Brea tar pits, located in Los Angeles. These are acre-size pools of black tar, also called pitch, asphalt, or bitumen. The pits contain tens of thousands of animals that became mired in the tar, including saber-toothed tigers, dire wolves, camels, mammoths, and birds, but no dinosaurs. The tar pits are post-flood formations, meaning they existed during a time in which dinosaurs were already scarce.

It must also be realized that some animal types simply were not fossilized. Consider two examples from recent history. A century ago, North American bison or buffalo numbered in the millions. They were largely eliminated by hunting and loss of habitat. However, no fossils of modern bison are found. Their bones simply disintegrated out in the open with no burial. Likewise, during the early 1800s the passenger pigeon was the most common bird in the Midwest. It far outnumbered all other bird species combined. However, hunting, disease, and habitat loss drove the passenger pigeon to extinction. The last survivor, a female named *Martha*, died in the Cincinnati Zoo in 1914. No known passenger pigeon fossils exist today.

In the evolutionary view the mammals arose millions of years after the dinosaurs. The mammal definition includes a warm-blooded nature and maternal suckling of the young. Mammals include, for example, bears, deer, whales, and also mankind. Note that this does not imply that mankind is in the animal category. According to Genesis 1–2 we are a distinct part of creation, with management responsibility over the animal world. In the creation view, all creatures originally lived together on the earth.

Many biologists now date modern mammals to 100 million years ago, well into the dinosaur era. Thus it can no longer be said—even by evolutionists—that dinosaurs and modern animals lived during entirely different periods of earth history.

27. Are there "modern cousins" of the dinosaurs?

In the evolutionary view all plants and animals are genetically related through a common ancestor. The creation view, in contrast, sees the many plant and animal kinds as distinctly made. Any similarities come from the handiwork of the Creator. There are several creatures living today that have an appearance, at least superficially, like dinosaurs or coexisting animals. Several will be described here.

For years there had been rumors of ferocious prehistoric beasts living on remote islands in Indonesia. In 1912 zoologists Van Hensbroek and Ouwens located the Komodo dragons there, the last of the giant lizards. In the monitor lizard family, these Komodo dragons weigh about 150 pounds and average 7–8 feet long. Lengths up to 30 feet have been claimed in past years. These reptiles have a poisonous bite and large claws. Females bury their eggs in deep trenches. The babies tunnel their way out, then spend their early years living in trees. The adults can move swiftly and have a dinosaur-like appearance. Iguanas, living in many tropical areas, also resemble dinosaurs on a smaller scale. This is not to say that modern lizards are simply small dinosaurs. The creatures differ substantially in gait and bone structure.

Aepyornis is a large extinct bird that lived in Madagascar about 2,000 years ago. The last of these elephant birds were hunted for food. This creature weighed up to a half-ton and was 10 feet tall. It looked quite similar to birdlike fossils recently found in China (see Question 9). The *Aepyornis* laid enormous eggs that were 15 inches long, with a volume equivalent to 180 hens' eggs.

The largest reptile in the world today is the estuarine or saltwater crocodile. It lives around the western Pacific Rim including southeast Asia, Vietnam, and Australia. Adult male crocodiles may reach a length of 23 feet and weigh well over 1,000 pounds. A particular captive crocodile in Thailand weighs 2,500 pounds (Matthews, 1998).

---

28. Describe the marine reptiles.

Three types of marine reptiles are recognized. These are the plesiosaurs, icthyosaurs, and the mosasaurs. The plesiosaurs were long-necked creatures whose fossils are found distributed worldwide. With paddlelike limbs and streamlined bodies they probably could easily swim forward, backward, or even upside down. One type of plesiosaur fossil called *Elasmosaurus,* "ribbon lizard," was found in a rock ridge along a river on Vancouver Island, British Columbia, in the 1990s. This creature was 43 feet long (13 meters), fully half of which consisted of head and neck, with large vertebrae 10 inches in diameter. The *Elasmosaurus* often may have swum at the surface with its head above the water.

Ichthyosaurs were streamlined, bullet-shaped swimmers. Sharks, dolphins, and ichthyosaurs are quite similar in appearance. However, since the first is a fish, the second a mammal, and the third a reptile, evolutionists see no close link between them. Some ichthyosaurs reached a length of 50 feet (15 meters). Engineers have studied the ideal shape for underwater animals and also for seagoing vessels. The theoretical form should allow for easy movement through water with minimum water disturbance or energy loss. The result from computer studies is a streamlined shape with a length 4.5 times the body diameter at its widest. The ichthyosaur length was about 5 times its maximum diameter, close to the ideal figure, which shows superb design (Alexander, 1989). Like tuna and dolphins, the ichthyosaurs also were probably warm-blooded, or endothermic. Their fossils indicate large eyes and also an ability to change the shape of the eye lens, similar to humans. This would result in extremely sharp underwater vision. Therefore the optics of ichthyosaurs clearly were not primitive evolutionary accidents. As with all other animals, an evolutionary explanation for the ichthyosaurs is lacking.

It is generally assumed in the evolutionary view that the ancestors of marine reptiles lived on land, then later moved back into the water. This same scenario is also applied to modern whales and dolphins. Somehow, land cattle moved into the sea and exchanged their legs and hair for flippers and blubber. It is a fantastic evolutionary story without supporting evidence.

Some German ichthyosaur fossils appear to have been preserved at the exact moment of birth. Babies are found partially removed from the mother. This would indicate very rapid, catastrophic burial. These unusual fossils alternately may show the spontaneous expulsion of the baby after the mother's death.

Mosasaurs resembled large seagoing lizards with flippers instead of legs. Crocodiles may have the closest modern appearance. One dinosaur-era mosasaur is called *Phobasuchus,* or "horror crocodile." This creature had a 6-foot skull and a body that was 50 feet long. It was truly a monster of the sea.

Turtles also inhabited the dinosaur world and were largely unchanged in appearance from today. Turtle fossils are found worldwide, and are especially numerous in Germany and Thailand. Some early sea turtles were as much as 13 feet (4 meters) in length, longer than a car. Smaller sea turtles survive today as marine reptiles.

---

29. Describe the flying reptiles.

The pterosaurs were a large group of flying reptiles living together with the dinosaurs. Evolutionists have had little success in explaining the origin of animal flight. They are forced to conclude that flight began on at least four separate, independent occasions in history. This includes the rise of the reptiles, mammals such as bats, the insects, and birds. Flying fish also might be added to this list since they use their pectoral fins as gliding wings. Regarding the origin of flying reptiles, one popular theory is that early thecodont reptiles first began airborne movement by hopping about in search of food. Flight then gradually resulted as scales evolved into feathers and the hops advanced to full flight. However, no evolutionary fossil link has been found between thecodonts and flying reptiles.

The first pterosaur was found in Kansas in 1870, and today is called a *Pteranodon,* meaning "winged" or "toothless." These creatures had leathery wings supported by a single long wing "finger" instead of four fingers like modern bats. Some of the fossil skin impressions also indicate the possibility of fur. The smaller flying reptiles, often called *Pterodactyls,* were long thought to be clumsy. Closer study has shown them to be precision-designed flying creatures. They grew lightweight hollow bones with a tubular design (figure 5). Some of their bones were also filled with small air sacs for added buoyancy.

Figure 5

Illustration of the hollow bone design of many pterosaurs. The figure includes bone deterioration that occurred after death.

Most pterosaurs were small, about the size of common birds. However, some flying creatures were impressively large. The largest known pterosaur fossil was found in Big Bend National Park, Texas, in the 1970s. Named *Quetzalcoatlus* after a feathered Aztec deity, this impressive flier had an estimated wingspan of 40–70 feet (12–21 meters) and weighed 200–350 pounds (Lauson, 1975). The large uncertainty in size results from the incomplete fossil remains that have been found. The complete fossil of a *Pteranodon* showed a 22-foot (7 meter) wingspan, twice the size of the largest modern bird, which is the albatross. *Pteranodons* may have soared above the sea on thermal air currents. Some of their fossils include fish scales and bones as stomach contents, probably from their final meal before death.

Models of the flying reptiles have been analyzed using wind tunnels and aerodynamic principles. The surprising result is that some of the larger specimens appear incapable of flight since they lack the necessary lift. However, they are clearly designed to fly. The solution to this problem may involve the vapor canopy that many creationists believe surrounded the early earth (see Question 4). The weight of this canopy of moisture would have increased the earth's atmospheric pressure to perhaps twice its present value. A greater air pressure would increase the buoyancy, and hence the soaring ability, of the large flying reptiles.

Fossils of *Archaeopteryx*, "ancient wing," have complicated the evolutionary theory of flight (see Question 9). This true bird is not thought to be even remotely connected with the flying reptiles. Another example, the *Teratorn*, was a large feathered bird from dinosaur times. Clearly, feathers did not first evolve at this time since they were already present. Another fossil specimen from Argentina, *Argentavis*, had a wingspan of about 20 feet (6 meters). Its individual, impressive feathers were probably 5 feet long and 1 foot wide.

In evolutionary thought the early birds had no complex songs and no migration ability. In contrast, the creation view maintains that birds have flown across the earth from the beginning of time. Scripture refers to the migration of the hawk (Job 39:26), and also the stork, dove, swift, and thrush (Jer. 8:7).

Figure 6

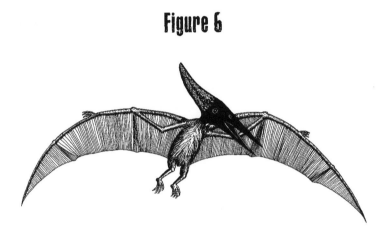

One of the pterosaurs, or flying reptiles.

Part 4

The Biology of Dinosaurs

30. Why were dinosaurs so large?

Not all dinosaurs were large, although the giant representatives get most of the publicity. Of the nearly 1,000 known distinct dinosaur species, the average adult size was that of a dog or sheep. For example, the grown *Compsognathus* weighed only 15 pounds, about the size of a large cat. The *Psittacosaurus* was smaller yet, similar in size to a squirrel. With that qualifier, however, the sauropod dinosaurs were truly impressive. They had a massive torso, pillarlike legs, and a long neck. The herbivore *Argentinosaurus* from Patagonia, Argentina, was 100 feet long and weighed 100 tons, half the weight of the blue whale. This dinosaur weighed more than a Boeing 727 airliner. The *Argentinosaurus* may have been the heaviest beast ever to walk on land. Also found in Argentina, the largest predator may have been *Gigantosaurus,* an economy-sized *Tyrannosaurus,* 42 feet long and weighing 8 tons (Schreeve, 1997). Table 6 gives the weights of several representative animals. The values given for dinosaurs are estimates based on their bone size and also on reconstructed models. Uncertainties in actual dinosaur body weight may reach as much as 50 percent due to incomplete fossil remains.

Table 6

The typical weights of representative adult animals, including dinosaurs, in order of decreasing size

| | Adult Weight | |
Animal	Pounds	Tons
Blue whale	400,000	200
Argentinosaurus	200,000	100
Brachiosaurus	160,000	80
Ultrasaurus	110,000	55
Seismosaurus	100,000	50
Apatosaurus	70,000	35
Supersaurus	60,000	30
Diplodocus	36,000	18
Tyrannosaurus	16,000	8
Triceratops	12,000	6
Stegosaurus	10,000	5
African elephant	10,000	5
Hippopotamus	5,000	2.5
Allosaurus	4,000	2
Polar bear	1,600	.8

Many reptiles grow throughout their lifetime. An example is the Nile crocodile, the largest of the living reptiles. An early, rapid growth rate slows gradually as the crocodile ages. Some mammals such as elephants also grow slowly throughout their lives. The largest dinosaur fossils therefore may belong to creatures that had lived for centuries. Two evidences suggest a rapid early growth rate for the dinosaurs. First, few half-grown dinosaur fossils have been found. Thus dinosaurs may have rapidly passed through their juvenile phase. It may also be that smaller, fragile bones simply did not often survive the burial and fossilization process. Second, dinosaur bone texture, especially for the stegosaurs, shows a rapid youthful growth. Robert Bakker estimates that stegosaurs may have grown from an egg to 5 tons in just one decade (Bakker, 1986).

The creation view sees purpose and design in nature, including the impressive size of some dinosaurs. The long necks of the sauropods may have allowed them to eat the foliage from tall trees. In this way dinosaur parents could reach a higher food source and avoid competition with their young. However, some paleontologists doubt that sauropods could extend their necks upward in this way. In general, large dinosaur size also provided defense against attack from smaller foes. In the animal world there is a measure of safety in large size. Predators rarely attack something that is big enough and strong enough to injure them instead.

There are physical limits to the maximum size creatures can attain because the supporting bones and muscles must increase greatly as animal weight increases. Consider elephant bones, which are so large that the legs themselves must be greatly thickened. Whales require smaller bones because of water buoyancy, but are in danger of fractured ribs if stranded out of water.

Suppose we double the size of an animal in all three dimensions including its length, height, and width. The animal's volume and weight will then increase 8 times (2^3). However, the bone strength only increases 4 times (2^2). This is because the bone's supporting strength depends on the cross-sectional area. The numerical comparison made here is between volume and area. Likewise, if we could make a creature 10 times larger, its weight would increase 1,000 times (10^3) while its bone strength becomes only 100 times greater (10^2). The implication is that, for large animals, bone size must increase disproportionately larger than body size. Galileo first wrote about this challenge to large size over three centuries ago, in 1638. This same area-volume reasoning also limits the possible size of trees. Perhaps you have seen the wide trunk needed to support a 350-foot redwood tree. The strength of a tree depends on its cross-sectional area, as exposed when the trunk is cut through. Table 7 gives some comparative bone diameters for animal sizes. Animals much larger than the dinosaurs would require bones too bulky to be practical. King Kong and Godzilla can exist only in movies, not in the real world.

Table 7

Relative size of animals
and their necessary supporting bones

Animal Size	Required Bone Diameter
1	1
2	2.8
3	5.2
4	8
5	11.2
6	14.7

Numbers in the second column are the $3/2$ power of the first column. Notice that bone size must increase substantially out of proportion to animal size.

31. Did dinosaurs have small brains?

In years past dinosaurs were commonly pictured as primitive, dim-witted, lumbering giants. They had a reputation for large muscles and fearsome teeth but little thinking ability. This view has changed completely as we have learned about their true complexity. Most large reptiles have relatively small brains, but this actually says little about their behavior or thinking ability. As researcher Glenn L. Jepson of Princeton University has written, a creature's brain is like a wallet: The contents are much more important than the size. Even a small brain is capable of complex processing ability. There is little correlation between intelligence and brain size in living creatures.

The weight of many animal brains is found to follow a power equation based on the animal's total weight. For birds and mammals, brain weight closely follows this relation:

$$\text{Brain weight} = .07 \, (\text{animal weight})^{2/3}$$

For fish and reptiles the relation is found to be different, giving a brain ten times smaller:

$$\text{Brain weight} = .007 \, (\text{animal weight})^{2/3}$$

Brain weight is also a direct measure of the brain's volume or size. Some animals' brains are found to exceed the calculated values, while others have a smaller than expected actual brain weight. One comparative measure of brain size is called the encephalization quotient (EQ), defined by paleontologist Harry J. Jerison in the 1960s. The EQ compares an animal's actual brain size with the prediction from the previous equations:

$$EQ = \frac{\text{Actual brain size}}{\text{Calculated brain size}}$$

Table 8 lists the EQ values for mankind and also for several animals. An EQ value less than one implies that the creature's brain size is smaller than expected. Values greater than one imply that the brain is larger than expected for this size creature. Note from the table that this unexpectedly larger brain size occurs in some dinosaur species. The conclusion is that not all dinosaurs had undersized brains. This is especially true for the *Stenonychosaurus* and *Allosaurus*. For dinosaurs the measured brain size is based on fossil skull measurements. Dinosaur brains were no smaller, relative to animal size, than modern lizard and crocodile brains. Therefore, dinosaurs were not dull-witted, primitive animals.

Table 8

The encephalization quotient, or EQ, for man and several animals

Creature	Encephalization Quotient (EQ)
Diplodocus, Brachiosaurus	0.1
Stegosaurus, Triceratops, Ankylosaurus	0.2
Protoceratops, Iguanodon	0.6
Tyrannosaurus	0.9
Elephant	1.2
Allosaurus	1.4
Man	1.5
Stenonychosaurus	5.3

Values greater than 1 imply that the creature's brain is larger than expected for a creature of its size.

Table 8 shows that stegosaurs had small brains, only walnut-sized in some cases. This does not mean they were "dumb" animals in any sense. However, it suggests that they probably relied on defensive armor rather than swift motion when they encountered predators. Incidentally, the largest known animal brain, reaching 20 pounds, belongs to the sperm whale. In comparison the human brain weighs about 3.5 pounds. The stegosaur's brain weighed 2–3 ounces.

---

32. Did some dinosaurs have two brains?

Stegosaurs and several other large dinosaurs have sometimes been credited with having two brains. An enlargement of the spinal chord in the hip region at the base of the tail was noticed in the 1800s and was called a *sacral brain*. The popular image was of a creature so large that a second brain was needed to help steer its back end, somewhat like a ladder fire truck turning a corner with the help of a real steering wheel at the back end as well as the front. Bert Leston Taylor of the *Chicago Tribune* further promoted the idea of two brains with a poem published in 1912 titled "The Dinosaur":

> Behold the mighty dinosaur
> Famous in prehistoric lore,
> Not only for his power and strength
> But for his intellectual length.
> You will observe by these remains
> The creature had two sets of brains—
> One in his head (the usual place),
> The other at his spinal base.
> Thus he could reason "A priori"
> As well as "A posteriori"
> No problem bothered him a bit
> He made both head and tail of it.
> So wise was he, so wise and solemn,
> Each thought filled just a spinal column.

If one brain found the pressure strong
It passed a few ideas along.
If something slipped his forward mind
'Twas rescued by the one behind.
And if in error he was caught
He had a saving afterthought.
As he thought twice before he spoke
He had no judgment to revoke.
Thus he could think without congestion
Upon both sides of every question.
Oh, gaze upon this model beast,
Defunct ten million years at least.

The sacral brain or reflex brain was actually an enlarged group of nerve cells. It was 20 times larger than the stegosaur's cranial brain. This ganglion probably was not used for actual thinking but instead was a routing system for nerve signals. A similar rear nerve concentration is common to many reptiles today. Ostriches also have a spinal cord enlargement inside their hip vertebra. For the stegosaurs this enlargement would have been useful in coordinating movements of the large tail. In this way the ladder fire truck analogy has some merit. However, the humorous idea that the stegosaur could have "thoughts and afterthoughts" with two complete independent brains is probably not correct.

---

33. Describe the skin of dinosaurs.

No actual dinosaur skin has been found, nor any of their flesh. However, detailed skin impressions exist where dinosaur hide was pressed into mud, which then hardened to stone. Some of these surface impressions look like elephant skin, while others look pebbly like the skin of a crocodile. A detailed fossil print of *Stegosaurus* skin was found in China in 1998. Figure 7 shows this interesting surface design with its starlike pattern. The bumps illustrated are the size of small coins. A very similar skin impression from a duck-billed dinosaur was also found in New Mexico.

The skin color of dinosaurs is unknown. Like many animals today, dinosaurs were probably multicolored with patches, bands, stripes, and spots to blend in with their surroundings. Alternately they may have displayed bright color markings to attract mates of their own species. The current lack of knowledge regarding dinosaur coloring has led to considerable speculation. For example, it is possible that some dinosaurs could undergo skin color changes depending on their surroundings, somewhat like a chameleon.

Figure 7

A drawing of a small portion of *Stegosaurus* skin.

---

34. Were dinosaurs healthy?

Some researchers have suggested that dinosaurs suffered from obesity, bone disease, and a host of other problems. Fossil studies, however, reveal just the opposite: Dinosaurs were generally very healthy animals. The study of ancient diseases and injuries is called *paleopathology*. The fossilized bones and teeth of dinosaurs provide the main evidence for their excellent health. The large sauropods do not show any obvious evidence of bone damage. Their impressive bones were fully sufficient to support their bulk. However, around 25 percent of the carnivorous theropod fossils show fractures of their forelimbs or feet, many of them healed (Farlow, 1997). This finding suggests a lifestyle of frequent violent struggle with prey, or perhaps competition with one another. Ribs from allosaur fossils sometimes also show healed fractures. Some of these injuries may have resulted from tripping and falling while running.

Dental abscesses are practically unknown in dinosaur fossils. A study of 10,000 fossils also found almost no bone disease such as osteoarthritis, a common ailment in older people (Farlow, 1997). Osteoarthritis is a bony overgrowth in joints, or cyst formation within bone. It appeared quite rarely in dinosaurs. Among thousands of fossils, just two iguanodon specimens in a Brussels, Belgium, museum show ankle bone overgrowth. Dinosaur bone joints seem to have been efficient and highly stabilized. Gout evidence has been found in rare cases, diagnosed from unusual lesions found on bone joints. A *T. rex* specimen in the Denver Museum of Natural History shows such lesions on its forelimbs. Gout disease results from elevated levels of uric acid in the blood, sometimes a dietary problem. Another single dinosaur bone from Cedar Mountain, Utah, also apparently had tumors within it.

It appears that dinosaurs as a whole did not suffer greatly from the common diseases that afflict modern animals. This would be consistent with a perfect creation, followed by a gradual increase in degeneration and disease following the Genesis 3 curse. Since dinosaurs lived closer to the beginning of time, they were less apt to have the physical mutation problems that have accumulated over history.

35. Describe dinosaur eggs and nests.

Many dinosaurs, if not all of them, were oviparous, or egg layers. In 1923, great numbers of petrified dinosaur eggs and shell fragments were found in the Gobi Desert of southern Mongolia. This remote region is one of the richest dinosaur nesting sites known. The eggs were found within nests full of vegetation that long ago had been gathered by the parents. These nests, on the ground, are bordered by a circular arrangement of stones (figure 8). As the plant material decayed it gave off heat to incubate the eggs. Clearly, dinosaurs were designed for success. Modern crocodiles sometimes warm their eggs in a similar fashion. Many of the dinosaur egg clusters are found deposited in a neat spiral pattern, totaling 30 or more eggs in each nest. Most of the dinosaur eggs are oblong and often about 8 inches in length. Some of the eggs contain fossilized embryos. The eggs have tiny airholes, like chicken eggs, which allowed the dinosaur embryos to breathe. The largest eggs, from sauropods, have a volume of nearly 3 quarts and may reach the size of bowling balls.

Dinosaur eggs have now been found at more than 200 sites worldwide, including the western United States. Paleontologist Jack Horner has found hundreds of hadrosaur dinosaur eggs in Montana, at a location called Egg Mountain. One nest holds 11 small dinosaurs, each about 3 feet (1 meter) long, all fossilized. Sauropod egg fragments are also numerous at southern locations in Argentina. Petrified embryos of sauropod titanosaurs have also been found there. In Portugal, three different types of dinosaur eggs were found in nearby nests. This particular site appears to have been a common nursery for several different dinosaur species.

In 1995, an ostrichlike dinosaur fossil 8 feet long was found apparently sitting upon 15 fossilized eggs. The *Ovi-*

raptor had its legs folded beneath its body, somewhat like that of a brooding mother hen. This unusual find was in the same remote area of the Gobi Desert where many other eggs are found. Investigators suggest that the parent dinosaur and eggs were buried by a flash flood of sand and mud. This particular fossil has heated up two ongoing debates. First, some experts see an evolutionary dinosaur-bird connection in this find because of the dinosaur's apparent brooding behavior.

Figure 8

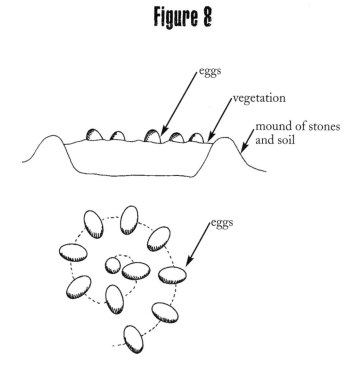

Drawing of a dinosaur nest with eggs, vegetation, and a border of stones. The nest is several feet in diameter. The eggs were often deposited in a neat spiral pattern.

Dissenters point out that reptile python snakes also lie on their eggs, and snakes definitely are not birds! Some experts also suggest that the squatting adult dinosaur from the Gobi Desert was possibly eating the eggs instead of protecting them. In fact, the very name *Oviraptor* means "egg-stealer." A second point is that if *Oviraptor* indeed sat upon its eggs to warm them, this action strongly suggests that dinosaurs were warm-blooded (Norell, 1995). This temperature debate is further discussed in Question 38.

The temperature of incubating dinosaur eggs may have determined their gender. This is the case for some reptiles. For alligators, a low temperature of 82–86 degrees Fahrenheit (28–30 degrees Celsius) yields females. A higher temperature of 90–93 degrees Fahrenheit (32–34 degrees Celsius) results in males. Gender is fixed during the second and third weeks of the alligator eggs' 8-week incubation period. Some crocodile species also produce females when the eggs are cooler. In other reptile species, temperature has the opposite effect on gender.

---

36. What was the diet of dinosaurs?

Dinosaur tooth structure is useful in determining their herbivore or carnivore diets. For example, the small *Stegosaurus* teeth clearly were designed for chewing vegetation, similar to cattle teeth. The vast majority of dinosaurs were probably plant eaters. In contrast, *Tyrannosaurus* teeth are shaped more like daggers and are as long as 9 inches.

Paleontologists have long wondered how the large herbivores could possibly consume enough food for survival. A full-grown elephant eats almost continuously, taking in 100–200 pounds of vegetation each day. The *Stegosaurus* was larger than an elephant, yet had a smaller head and mouth. There must have been rich leafy plant sources available for these large plant-eating dinosaurs. One discovery of possible stomach contents in a herbivore was reported in the early 1900s. At this time, conifer needles, twigs, and seeds were found in the body cavity area of an *Edmontosaurus* fossil (Farlow, 1997).

Some of the carnivores also may have satisfied their hunger in a way similar to the boa constrictor, which can swallow large prey. Skull fossils of *Ceratosaurus* and *Allosaurus* show a loosely hinged jaw that could be dislocated. In this way they may have been able to enlarge their mouths considerably to take in large portions of prey or vegetation at one time.

The creation view gives valuable insight concerning dinosaur diets. Genesis 1:29–30 declares that originally all animals, as well as people, were plant eaters. Thus the early world was entirely vegetarian in nature. After the great flood, Genesis 9:3 explains that people were allowed to eat meat. Some of the animal world also became carnivorous either at this time, or perhaps earlier at the time of the curse of Genesis 3. Secular science alone has not arrived at the conclusion that all animals underwent a major dietary change in the past. The Bible here suggests a possible area of creation research: How were animals able to adjust to this change? Was there a supernatural adjustment of their tooth structure and internal metabolism? Could the animal world someday revert back to the original vegetarian state? The answer to this last question is a definite yes. In God's plan, all animals will eventually return to a nonagressive nature. This shift, described in Isaiah 11:6–7, will occur during a future millennial period:

> The wolf will live with the lamb,
>> the leopard will lie down with the goat,
> the calf and the lion and the yearling together;
>> and a little child will lead them.
> The cow will feed with the bear,
> their young will lie down together,
>> and the lion will eat straw like the ox.

The dietary pattern of living creatures appears to move in a full cycle, from plants to meat, then back to plants, as summarized in table 9.

Table 9

Bible references indicating the changing diets of mankind and animals during history

Diet	Mankind	Animals
Plants in the past	Genesis 1:29	Gen. 1:30
Meat at present	Genesis 9:3 Postflood	Began at the curse or postflood time
Plants in the future	Isaiah 11:6–9	

37. What are coprolites and gizzard stones?

Coprolites are fossilized animal droppings. The word comes from the Greek *kopros,* for dung. This topic may sound rather disgusting, but mineralized fecal remains provide important clues to animal diets in the distant past. For herbivores, the dinosaur dung is usually a rounded mass and varies from pebble- to basketball-sized. Herbivore droppings with broken-up pieces of conifers have been found at a site in Montana called the Two Medicine Formation. These particular remains are honeycombed with burrows made by dung beetles that scavenged the droppings while they were still fresh. The dinosaur-era beetles conflict with the usual evolutionary story, in which dung-eating beetles evolve long after dinosaur extinction (Travis, 1996). Coprolites from carnivorous or omnivorous dinosaurs are rarely found. A fossilized dropping found in Canada in 1997 contained 200 small bone fragments from other dinosaurs. Many birds ingest grit, which remains in their gizzards. These small pebbles, called *gastroliths*, aid in grinding up seeds before digestion. Alligators and crocodiles also swallow hard stones, often pieces of quartz or granite, to line their gizzards. It appears that the sauropod dinosaurs also swallowed egg-sized stones and retained them for digestion. Stomach stones are often found within the fossilized abdominal cavities of these large plant-eating dinosaurs. The churning motions of these stones helped pulverize the large quantity of branches and leaves in the dinosaur diet. Chemical effects from digestive acids generally give gastrolith stones a distinctive high polish. A single *Seismosaurus* excavation in New Mexico recovered more than 240 gastroliths, the largest specimen 4 inches (10 cm) in size. The marine reptiles may also have swallowed stomach stones to increase their weight. The resulting reduced buoyancy would have helped them control their movement underwater. Some fossil assemblages contain hundreds of pounds of such ballast rocks.

---

38. Were dinosaurs cold- or warm-blooded?

This question, along with what caused dinosaur extinction, is one of the greatest controversies in dinosaur studies today. The older, traditional view is that all dinosaurs were sluggish, cold-blooded reptiles. Also called ectotherms, or "heated from without," such animals regulate their body temperature by using sunlight and shade. In recent years some paleontologists have suggested an alternative view— perhaps dinosaurs had an active, warm-blooded nature. Such creatures, called endotherms, meaning "heated from within," maintain a constant body temperature by generating their own internal heat. In other words, warm-blooded creatures have internal thermostats with a particular setting. Fish, amphibians, and most reptiles are ectotherms, while birds and mammals are endothermic. Much of the current debate over dinosaur temperature results from evolutionary assumptions. For those who believe that dinosaurs became birds, a warm-blooded nature for dinosaurs is attractive. Paleontologist John T. Ostrom believes that the carnivorous theropod dinosaurs were warm-blooded while the giant herbivorous sauropods were not. The creation position includes the possibility of either warm- or cold-blooded dinosaurs, both kinds simultaneously, or perhaps a thermal nature in between the two extremes.

The following points outline some of the evidence for warm-blooded, endothermic dinosaurs.

1. Many dinosaurs had skeletal features that enabled them to walk at the typical speeds of warm-blooded animals, or even to run rapidly. Certain dinosaur trackways also suggest high-speed locomotion. Such a high level of activity and energy suggests warm-bloodedness.

2. Dinosaur bone texture reveals rapid growth, unlike modern cold-blooded animals. This rapid bone development may imply a high metabolism with an accompanying high internal temperature (Horner, 1988). However, cold-blooded crocodiles raised under very warm conditions also show rapid bone growth. Therefore, dinosaur growth may have had more to do with climate than with internal thermal regulation.

3. Dinosaur bone structure is more porous than that of most modern reptiles. This porosity indicates a high rate of blood flow, characteristic of warm-blooded animals. Cranial evidence also indicates a high level

of blood circulation in the brain and therefore an energetic body.

4. Modern warm-blooded predators are typically few in number when compared with populations of their prey. Modern predator-prey ratios are often as low as 1–4 percent. For example, rabbits in the wild greatly outnumber foxes or coyotes. Paleontologists find a similar low ratio for the number of fossil dinosaur carnivores such as allosaurs (Bakker, 1986), compared with the herbivore dinosaurs. These fossil counts may either represent actual populations or different fossilization rates.

5. Stegosaurs had vertical triangular plates along their spines that radiated away excess body heat. This anatomy suggests warm-bloodedness. The large sail-like back fins of *Dimatrodon* held a rich web of blood vessels, which likewise served as radiators. These fins could also collect sun energy for warming purposes. A major unknown is the amount of surface moisture that dinosaurs evaporated away in heat regulation. Evaporation is a major cooling mechanism for many creatures, including people.

6. Dinosaur fossils have been found in the Arctic and Antarctic regions. Accompanying fossil plants and geochemistry indicate a cool climate at that time with evidence of frost and frozen lakes. Cold-blooded creatures such as crocodiles could not cope with such low temperatures today. The polar dinosaurs may therefore have been warm-blooded to endure the challenging climate.

7. In certain cases, scientists are able to measure different varieties of oxygen atoms, called isotopes, within dinosaur bones. This isotope variation is known to indicate temperature differences at the time of bone formation. However, very little isotopic variation is mea-

sured between dinosaur body regions. This finding hints at little temperature difference between dinosaur extremities and deep body regions, a further characteristic of warm-blooded animals. Critics counter that later groundwater contamination of fossil bones may be responsible for the measured oxygen isotope values (Farlow, 1981). Every argument for a warm-blooded nature seems to have a counterargument.

There are also several lines of evidence for cold-blooded, ectothermic dinosaurs:

1. The obvious dinosaur comparison is with modern reptiles, all of which are cold-blooded ectotherms.
2. Modern endotherms have nasal structures called respiratory turbinates. These membranes moisten and warm the incoming air and also recover some heat from outgoing air. Dinosaur skull fossils seem to lack a space for this turbinate cartilage.
3. A well-preserved *Sinosauropteryx* fossil from China has revealed details about its lungs. These appear to be crocodile-like, incapable of achieving the high rate of air exchange needed by warm-blooded animals. This is countered, however, by a baby *Scipionyx,* the only dinosaur found to date in Italy. This fossil shows evidence of a diaphragm for breathing, similar to humans and other mammals.
4. A warm-blooded creature consumes 10 times as many calories as a cold-blooded animal of the same size. Large herbivore dinosaurs, if warm-blooded, would have faced a challenge to consume enough food to survive (Bakker, 1986).
5. No dinosaurs have been found with a layer of insulating feathers. Such a covering would not necessarily link dinosaurs with birds, but feathers would be evidence of endothermic temperature regulation.

In conclusion, perhaps both cold- and warm-blooded dinosaurs existed. Or maybe they were intermediate between the cold- and warm-blooded extremes. The large dinosaurs may also have maintained a nearly constant body temperature by *mass* or *inertia homeothermy*. In this view their bulk provided a large reservoir for heat. In other words, they may have kept warm simply by being large. This is the case for the large saltwater crocodiles of northern Australia. Although cold-blooded, their internal temperature has been measured to stay within a 2-degree-Celsius range day and night.

39. Has dinosaur DNA been found?

This idea was popularized in the 1990s book and movie titled *Jurassic Park*. The story plot involves mosquitoes that long ago ingested dinosaur blood. The insects were then entombed in amber. Dinosaur DNA was later extracted from these insect fossils to reconstruct living, prehistoric creatures. DNA is deoxyribonucleic acid, a double-helical molecule that carries genetic information.

No uncontested dinosaur DNA has yet been found in nature. Dinosaur fossil evidence is largely limited to their fossilized bones, teeth, and footprints. Fragments of dinosaur DNA may eventually be found, and there have been reports of dinosaur blood cells being discovered. This possibility indeed fits the creation view that dinosaurs lived until relatively recent times, with actual remains still possibly existing. However, the overall theme of the *Jurassic Park* movie is not likely. First, one would need complete dinosaur DNA molecules. However, such molecules for any plant or animal are very complex and fragile. Each molecule contains more than a billion carefully arranged atoms. The Second Law of Thermodynamics assures that such molecules disintegrate quickly over time if not maintained by a living system. DNA recovered from animal tissue just a few centuries old is highly degraded. Other studies show that DNA begins to degenerate within just hours of an organism's death. Even if dinosaur DNA were somehow available, the animal could not be simply produced. This genetic material must be nurtured within a compatible host animal, most suitably a living dinosaur! The Creator may choose to bring back the dinosaurs in some future age. However, it does not appear possible within the realm of present genetic research.

If not dinosaurs, could woolly mammoths ever be brought back to life? Along with mineralized fossils, frozen mammoth flesh is sometimes found in Arctic regions, especially northern Siberia. Suitably preserved mammoth sperm may one day be isolated. A modern female elephant could then be fertilized to produce a hybrid animal with some mammoth characteristics.

Woolly mammoths are said to have lived during the *Pleistocene epoch*, extending from 1.8 million years ago to relatively recent times. On Wrangel Island north of Siberia, mammoth fossils are dated at just 3,800 years old. In the evolutionary view, of course, all mammoths lived millions of years after the dinosaurs were gone. In the creation view the mammoths and dinosaurs likely died out together in the postflood era.

Part Five

The Physics of Dinosaurs

40. How do fossils turn to stone?

The word *fossil* comes from a Latin word meaning "dug up." A fossil may be defined as any natural record of life from the past. It may be a footprint, a bone, or petrified wood. Usually fossils are found as hardened minerals. The study of the process by which organisms are fossilized is called *taphonomy*. Fossilization is considered to be a rare occurrence in nature. A typical estimate is that only one in a few million deaths leads to a permanent fossil record. In addition, a rule called *Durham's Law* further estimates that only 10 percent or less of actual plant or animal species are ever preserved in this way. This would imply that we do not know about many organisms from the past, which may be true. Paleontologists estimate that they have found fossils of fewer than one out of every thousand animal and plant species.

Several steps are required for fossilization to occur.

1. A plant or animal must die and be buried very quickly. Otherwise decay or scavengers will quickly consume the organism.
2. Burial must continue long enough for the fossilization process to convert the organism to stone. Beneath layers of mud and gravel, circulating groundwater with dissolved minerals is needed.
3. Erosion is necessary at a later time so that the fossil can be exposed and eventually discovered.

Notice that all three of these steps fit the description of the global Genesis flood. This event may be responsible for most of the fossils known today.

Fossilization almost always involves *permineralization,* or simply mineralization. This describes the infusion of chemicals like calcium, iron, and silica into spaces within buried bones. Water percolating through the ground commonly carries these elements. Sometimes the original chemicals of the bones remain. In other cases a complete replacement of atoms takes place. Similar processes may change soil to sedimentary rock such as limestone, sandstone, and shale. Fossil samples also include natural casts. When an organism dies it may disintegrate and leave behind a mold, or hollow space. This cavity may later be filled with sand or mud, which hardens to rock in the exact shape of the original organism. This resulting type of fossil is called a cast.

The time needed for fossilization to occur depends strongly on the mineral content of the surrounding groundwater. The usual assumption of a multimillion-year requirement has been challenged by recent studies. There are many examples of objects being fossilized very quickly. For example, the waters of Lake Turkana in Kenya are very alkaline, containing a heavy concentration of various mineral salts. Porous objects lying near the lakeshore will turn to rock in less than a year by soaking up calcium carbonate from water in the soil. This rapid fossil conversion has been observed in cloth fragments and other lakeside debris.

Fossilization sometimes exactly records very minor details. A baby *Scipionyx samniticus* dinosaur, just 9 inches long, was found within the limestone rock of Italy in 1980. The baby's stomach is visible, as are tiny chest muscles. Even the tiny nails on its claws are preserved.

In special cases, fossils may exist as actual remains without mineralization. For example, some woolly mammoths are preserved in permafrost soil in a frozen state. Their actual skin and flesh is thus available for study. Mummified remains of sloths also have been found in dry caves in Nevada. Egyptian mummies could also be included in this unusual fossil category.

---

41. Do fossils support evolution?

If one is looking for evidence in support of evolution, the fossil record should be avoided. Instead of revealing numerous connecting links and gradually improving life-forms, fossils are a strong testimony to supernatural creation. The following five conclusions may be drawn from the fossils found all around the earth.

1. Creative design is seen at every level of life. One cannot accurately say that early plants or animals are less complex. For example, it is commonly taught that trilobite fossils are ancient ancestors of modern life. However, trilobite fossils reveal compound eyes that are even more "advanced" than our human eyes. All living things also contain cells, tiny "factories" whose structure and functions fill entire books with technical discussion. Every life-form has been created uni-

quely complex and designed from the very beginning of time.

2. Most fossils, including dinosaurs, formed rapidly during the worldwide catastrophe of the Genesis flood. It therefore should be no surprise that many fossils around the world are scattered, broken, and lodged in hardened mud. Fossils are a vivid reminder of the consequences of sin. God is patient with mankind, but judgment eventually comes. Fossils form only rarely today, yet there are billions in the ground. We walk upon a vast fossil graveyard.

3. Fossils show that plants and animals have remained in separate categories or kinds since the beginning of time. No uncontested missing link has ever been found in the fossil record. In his 1859 work *The Origin of Species*, Charles Darwin said very little about fossils because of the incompleteness of the fossil record. And today, with many more species known, there actually are more links missing than in Darwin's day. Also, many classic transition examples, such as the evolution of the horse, have been discarded or modified in recent years.

4. All efforts to relate one fossil to another on the basis of gradual change have been arbitrary, and often temporary, as evolutionary ideas change. Similarities, called *homology*, whether observed among plants or animals, show the common Creator rather than a common ancestry.

5. Fossils of man's alleged "ancestors" always fall into one of two categories. They are either fully human or fully apelike. No ape-man has ever been found in the fossil record, nor ever will be. One well-known type of fossil is *Australopithicus*, or "southern ape." Perhaps the best-known example is "Lucy," found in Ethiopia in 1974. *Australopithicus* is generally recognized as a

primate, an animal that could not easily walk upright. Some researchers argue, somewhat hopefully, that Lucy *did* walk upright, and was a "successful animal." Another fossil category is *Neanderthal.* These probably were rugged cave dwellers from Old Testament times. They used tools and fire, and had larger brains than modern man. Ceremonial flowers are sometimes found at their burial sites. The *Neanderthals* appear to have been fully human.

---

42. Describe the biomechanics of dinosaurs.

Biomechanics concerns the motions of living creatures and the resulting forces within their body structure. This includes muscle activity, speed of locomotion, and the strength of bone. All plants and animals show intricate design in their makeup. Consider, for example, a tree that supports its heavy branches nearly horizontally. The internal forces can be immense, far greater than the actual weight of the wood itself. If you have held your arm out straight for a period of time, you surely have experienced the fatigue that soon results. And yet a tree may last for centuries, continually becoming heavier and stronger. The following list illustrates some of the engineering principles that are implied from dinosaur fossils.

The camarasaur was a large sauropod dinosaur. Its large backbone vertebrae have an open, symmetrical, anchor-shaped structure (figure 9a). This lightweight design shows an economy of material, combining both open space and great strength. The structure supports the dinosaur's bulk somewhat like a steel I-beam supports a building. Its skull was also built in lightweight fashion with large openings (figure 9c). The camarasaur neck structure appears to be designed for lateral, side-to-side sweeping rather than for vertical motion. The dinosaur bone material itself was fully as strong as that of large modern animals.

Figure 9

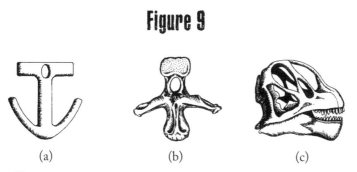

(a)   (b)   (c)

The symmetrica, lightweight backbone vertebra of camarasaur (a). Also shown is a vertebra for another, unknown dinosaur (b). The camarasaur skull was designed with openings to reduce weight (c).

Many of the large sauropods like *Diplodocus* had thick ligaments running the length of their backs, somewhat like cables on a suspension bridge. These elastic ligaments gave support to both the neck and tail (figure 10). With its body fully extended, these muscle ligaments would have experienced a tension of at least 30,000 pounds, or 15 tons.

Figure 10

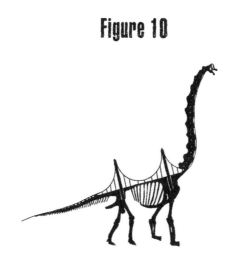

The back ligaments of *Diplodocus* were somewhat like the strong cables of a suspension bridge, shown here in comic form.

The *Tyrannosaurus* had large, very strong jaw muscles called *masseters*. Scientists have simulated bite marks found in bones by using replica *T. rex* teeth and a hydraulic press. The results show a biting force of at least 2,800 pounds. This jaw strength far outperforms wolves, lions, or sharks, and is in the same league with the largest crocodiles. The masseters also are the strongest muscles in the human body, normally exerting a maximum force of about 150 pounds, nearly 20 times less than those of *T. rex*.

Many theropods, including *T. rex*, probably moved about somewhat like animated seesaws. The large tail counterbalanced the weight of the head. The fulcrum, or pivot point, was positioned over the hips (figure 11). All of the balancing mechanisms of the inner ear were obviously created in place for these animals.

Figure 11

The overall balance of the *Tyrannosaurus*, pivoting on its legs.

Archaeopteryx is a pigeon-sized bird thought to have lived 150 million years ago. Seven fossil specimens have been found, all in Germany. There has been much debate as to whether *Archaeopteryx* is a "birdlike reptile" or a "reptilelike bird." Its wing structure declares that it is neither, but is instead a modern bird in every way. Although not a dinosaur, *Archaeopteryx* with its wing design shows the created design that existed from the very beginning of biblical history.

Figure 12 shows a magnified view of a common feather similar to that from an *Archaeopteryx*. Notice the tiny hooks on the side barbs. These hooks interconnect to produce a surface that readily sheds water and also provides a smooth wing surface for flight. These hook connections are very similar to the common Velcro fastener. Velcro has been listed as one of the outstanding modern inventions. Actually it has been present since the fifth day of creation, when all the birds were created, including *Archaeopteryx* (Gen. 1:20–21).

Figure 12

Magnification of a feather, showing its Velcro-like connections.

Serrated teeth are an additional example of functional design. Many of the theropods such as the *Tyrannosaurus* displayed this feature. Such teeth have a series of many small projections somewhat like the teeth of a saw (figure 13). This structure greatly helps the strength and cutting ability of teeth.

Figure 13

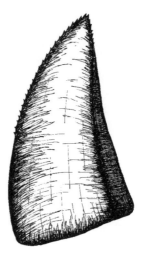

An illustration of intricate tooth serration, shown by dinosaurs such as *Tyrannosaurus.*

Design in nature is one of the older, historical arguments for creation. It was popularized in 1802, when William Paley (1743–1805) wrote the book *Natural Theology*. Design evidence has never been adequately refuted by skeptics of creation. It is referenced in Romans 1:20: "For since the creation of the world God's invisible qualities—his eternal power and divine nature—have been clearly seen, being understood from what has been made, so that men are without excuse."

43. What do we learn from dinosaur footprints?

The study of footprints and other trace fossils is called *ichnology*. This word comes from the Greek *ichnos,* meaning "trace." Dinosaur tracks are found worldwide by the millions if not billions. These prints are of special interest because, unlike bones, footprints give us information about the behavior of the creatures while they were alive and active. Dinosaur prints were first noticed in Connecticut in 1800. The strange tracks were studied by scientists from Harvard and Yale, and were said to be from "Noah's Raven," apparently a reference to Genesis 8:7. However, a raven does not produce tracks of dinosaur size! Dinosaur bones were later found in this same area.

Fossil footprints occur when tracks are made in mud or other sediment, then become hardened to permanent stone. Many of the uncovered dinosaur prints look very fresh, as if they were made just yesterday. For further study, casts are made from the impressions by paleontologists. First, the inner rock depression is coated with grease or silicon spray. Then plaster of paris is poured into the track. When dry, the hardened plaster cast can be pried loose as a permanent record. Latex rubber can also be used instead of plaster. Natural casts may occur if a track is later filled with sediment that then hardens.

Information gained from studying dinosaur tracks reveals much about the following points: First, footprint measurements help determine the locomotion and speed of dinosaurs (see Question 45). Unusual prints from an unidentified dinosaur in France indicate that it even may have hopped around like a kangaroo!

Second, footprints help us understand the morphology (size, shape, etc.) of animals. For example, the position of their prints shows that dinosaurs did not have a sprawling gait like modern lizards and crocodiles whose body positions look similar to a person doing push-ups. Instead, the dinosaur feet were positioned narrowly beneath the body, so their stance was much like an elephant's. Museums still show dinosaurs both ways, either with sprawling or vertical feet positions. Bipedal dinosaur tracks of theropods are quite narrow, with one foot in front of the other as expected.

Third, linear fossil marks from tail dragging are very rare. From this observation it is thought that tails usually were held outward, above the ground, and used for balance.

Fourth, the shape of their footprints helps distinguish between dinosaur types. Theropods made a three-toed print as they walked upright on two legs. In contrast, the large sauropods had broad feet with rounded toes. They walked on all four feet, and their hind feet were somewhat larger in size. Some of these impressive prints are 3 times the size of elephant prints. Usually the sauropod prints are indistinct and do not reveal the particular species of dinosaur.

Fifth, dinosaur footprints are occasionally found in great numbers. Tracks are so abundant in certain Colorado rock layers that the areas are called "Dinosaur freeways." The churned up, highly trampled ground is said to have undergone *dinoturbation*. These crowded mega-tracksites indicate that the animals may have traveled in vast herds. Perhaps the animals had far-reaching, seasonal migration routes.

At the Davenport Ranch in Texas, trackways show herds of apatosaurs moving together. Smaller tracks also appear in the center region, as if the young dinosaurs were being protected by large bulls on the perimeter of the group. There are also allosaur tracks in the vicinity. These carnivores may have been stalking the apatosaur herd, similar to the behavior of some modern hunting animals.

44. Are dinosaur and human tracks found together?

There have been many reports of the footprints of people and dinosaurs being in the same sedimentary rock layers. This evidence has been found in such places as Utah, Australia, and Russia. Dual tracks also occur in the limestone beds along the Paluxy River near Glen Rose, Texas, south of Dallas-Fort Worth. During the 1930s depression years some of these tracks, very humanlike in appearance, were chiseled out of the bedrock and sold to collectors by local residents. The Texas tracks were popularized by researcher Roland T. Bird in the 1940s. If genuine, these tracks are an insoluble problem for evolution. After all, dinosaurs supposedly disappeared 65 million years before mankind evolved and walked upon the earth.

It remains uncertain whether the observed humanlike tracks actually come from man or beast. One major complication is that a fresh dinosaur footprint disturbs several lower layers of mud (see figure 14). Much later, as fossil tracks are eroded, the lower impressions, called undertracks or "ghost tracks," may be exposed. When made by three-toed bipedal dinosaurs, either the upper or lower tracks may be elongated, similar to human footprints. The outer portions of the track may be shallow and thus erode away, leaving only the center, moccasin-like portion. In some other cases, claw traces show in the underprints but not in the initial layer on which the animal walked (Padian, 1986). This may be the reason why some Texas fossil prints appear to develop claws as they continually erode. Adding to the complication, one cannot readily distinguish between surface tracks and undertracks. Continued study will shed further light on the authenticity of these humanlike tracks. They remain a possible, fundamental conflict with the evolutionary timescale of history.

Figure 14

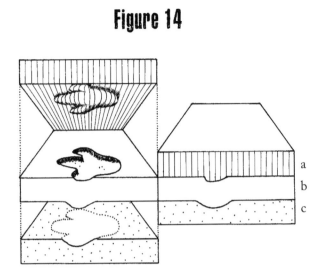

a
b
c

A preserved dinosaur track, showing the original track (a), a natural cast or mud infilling (b), and a deeper undertrack impression (c).

---

45. How fast did dinosaurs run?

There are at least three ways to explore dinosaur speed. First, one can measure the actual speeds of somewhat similar animals living today. Second, anatomical studies of dinosaur fossils give theoretical limits on their possible speed. Computer modeling of dinosaur movement is also helpful. Third, dinosaur footprints preserved in sedimentary stone give a permanent picture of their locomotion. Dinosaur *stride* is defined as the distance between two prints made by the same foot. *Pace* is the distance between successive footprints of two rear feet or front feet. Scientists have constructed dinosaur speed formulas based on the stride measurements combined with size estimates for the animal (Farlow, 1981). Fossil footprints cannot reveal whether the creature was moving at maximum speed, but some of the calculated values of speed are surprisingly high. Table 10 lists the top estimated speeds for several creatures, including people. No footprint fossils that have been found to date show the large elephant-type sauropod dinosaurs moving rapidly. Instead, their typical speeds are similar to human walking speed, about 2–3 feet per second.

Table 10

Estimated top speeds of various creatures, including several dinosaurs and also people, in order of increasing values

Organism	Top Speed	
	meters/ second	miles/hour
Stegosaurus	2–17	4–38
Apatosaurus	3–12	27–75
Brachiosaurus	5–12	11–27
Tyrannosaurus	6–20	14–43
Triceratops	7–12.5	16–28
Penguin (swimming)	7	17
Person	10–12	25
Kangaroo	14	31
Killer Whale	15	35
Deinonychus	12–25	26–55
Dromiceiomimus	14–31	32–68
Ostrich	14–18	36
Greyhound, Horse	20	45
Lion	22	48–50
Pronghorn Antelope	24	55
Cheetah	30	70

It appears that the *Tyrannosaurus* had to be especially careful in its movements. Its small forearms would have provided little protection in a sudden fall. Severe head injury could have easily resulted if the creature tripped while running. Therefore, the *T. rex* may have moved about slowly, choosing its steps with caution. The same hazard applies today to giraffes, which will likely break bones in a fall. Sensing this, they display a slow, graceful movement.

---

46. Describe dinosaur temperature regulation.

Animals differ greatly in their ability to maintain body temperature. In general, small animals have relatively more surface area to radiate away heat, and large animals have relatively less area. As a result, small animals experience more heat loss than large animals, relative to their size. Therefore, birds have feathers, and chipmunks are provided with fur for heat insulation. Even so, small creatures must eat almost constantly to make up for heat loss. If birds or rodents hibernated like larger creatures, heat loss would be fatal, and they would never wake up. At the other end of the size scale, large animals must get rid of excess body heat, and they are designed with this need in mind. An elephant, for example, has loose skin for increased surface area. It needs no fur insulation, and the large trunk and ears also radiate heat. Large dinosaurs, like elephants, also were designed to dissipate extra calories. Whether cold- or warm-blooded, the large dinosaurs were equipped with heat radiators. *Stegosaurus* had multiple protruding back plates. These plates held a rich web of blood vessels. Exposed to the air, they served as thermoregulators, both dissipating heat and, at other times, also gathering warmth from the sun. The *Dimetrodon's* large back sail served a similar purpose. For the large sauropod beasts, long necks and tails provided additional surface area for heat radiation.

It was the large dinosaurs that had "heat to spare." It is not surprising, therefore, that dinosaur fossils are found in polar areas. A *Pachyrhinosaur* skull was found on the North Slope of Alaska, just 350 miles from the North Pole (Bakker, 1986). Dinosaur bones are also found in southern Australia, at 70 degrees south latitude. The fossilized plants and geochemistry imply that this region was cold with frost and frozen lakes during the dinosaur era.

Figure 15

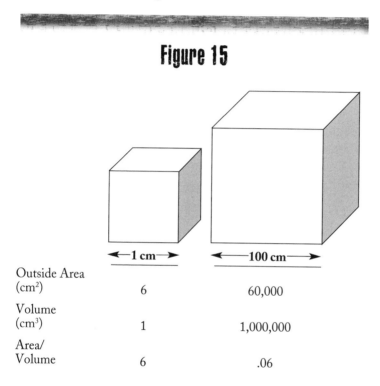

	←1 cm→	←100 cm→
Outside Area (cm²)	6	60,000
Volume (cm³)	1	1,000,000
Area/ Volume	6	.06

An area-volume comparison between small and large cubes (not to scale). These cubes represent small and large animals.

As an analogy for small and large animal size, consider the two cubes shown in figure 15. Suppose the cube sides measure 1 centimeter and 100 centimeters respectively. The outside area of the cubes represents the animal's exposed skin. As the figure shows, the large cube naturally has a much greater surface area. Next consider the cubic volumes, representing the actual bulk or weight of the animal. Naturally the large cube again has a much larger volume. Now divide the area by the volume for each cube. This ratio is found to be 100 times greater for the small cube. This implies that a small animal, relative to its size, has more surface area and loses more heat than a larger animal. The conclusion is that large dinosaurs would have had little trouble staying warm. Their greater challenge may have been getting rid of excess body heat.

47. How would increased air pressure affect dinosaurs?

One popular option in the creation model is that the early earth was surrounded by a vapor canopy. In the form of invisible water vapor, this canopy could have held the equivalent of many feet of water. It would have been located in the upper atmosphere, called the stratosphere, 10–15 miles high. From the time of Adam to Noah, this canopy gave the entire earth a warmer climate than exists today. The moisture layer resulted in an ideal "greenhouse effect" all over the earth. The canopy is no longer present since it collapsed during the Genesis flood, providing one source for the floodwaters. The other source was groundwater (Gen. 7:11).

If such a canopy indeed existed, its weight would have more than doubled the earth's present atmospheric pressure (Dillow, 1981). This greater air pressure would have profound implications on the earth. It may be a partial reason for the longevity of Old Testament patriarchs. Old Testament genealogies reveal that the generations from Adam to Noah lived for many centuries (Genesis 5). Modern studies have shown advantages to a hyperbaric, or high-pressure, environment for certain medical procedures. These include treatment for bone infections, complication from radiotherapy, and chronic, nonhealing wounds.

A higher atmospheric pressure also would have affected the animal world. For example, it may have helped the respiration of large creatures such as dinosaurs. Consider also *Quetzalcoatlus*, a pterodactyl fossil found in west Texas. This was a large flying reptile with a wingspan of 40–50 feet. Research by paleobiologists has indicated that *Quetzalcoatlus* would have needed an atmospheric pressure at least twice the present value to be able to fly. If it was alive in the world today, the *Quetzalcoatlus* simply would not be able to fly. This may be further evidence for a vapor canopy during the preflood era, when this magnificent creature surely flew overhead.

---

48. Could some dinosaurs breathe fire?

This unusual question arises from Job 41, which describes a marine reptile called leviathan (see Question 14). Here are some descriptive phrases for the leviathan:

His snorting throws out flashes of light (v. 18).
Firebrands stream from his mouth; sparks of fire shoot out (v. 19).
Smoke pours from his nostrils (v. 20).
His breath sets coals ablaze (v. 21).
Behind him he leaves a glistening wake (v. 32).
Nothing on earth is his equal (v. 33).

Many scholars have concluded that the creature described in Job 41 cannot be real. Instead, leviathan must simply be a mythical exaggeration of some common animal. However, the detailed biblical description of leviathan appears to be very literal.

Whether or not the living creature called leviathan actually breathed fire is a separate question. Some animals can expel respiratory exhaust that appears much like smoke. And an aquatic animal can stir up luminescent light. Such luminescence often is caused by biochemical activity of microorganisms.

On the other hand, a fire-breathing nature is certainly physically possible. Let's consider several points. First, any creature is capable of internally producing flammable methane gas. It is a natural product of food breakdown. This gas could then be exhaled and perhaps ignited by a spark. We know that electric eels are capable of producing 500 volts of static electricity. With the right physiological arrangement, this voltage could easily produce a spark.

Second, an alternative approach would be the internal mixing of chemicals to produce an explosive reaction. A famous example is the bombardier beetle. For defense, this small creature readily produces a directed stream of noxious gas with a temperature of 212 degrees Fahrenheit (100 degrees Celsius), the boiling point of water. Many dinosaurs had skull chambers and hollow crests that could possibly have been reservoirs for component chemicals.

Third, consider the many traditional stories from the past that describe fiery dragons. Certain dinosaurs may be the origin of such stories.

---

49. What sounds did dinosaurs make?

Beyond the typical growls and snorts of all animals, the dinosaurs with headcrests may have had unique sound abilities. These crests were hollow and apparently served as resonating sound pipes, similar to a large flute or trumpet. As one example, the duckbill dinosaur named *Parasaurolophus* had multiple air passages in its crest. Starting at the nose, air moved up through the bony crest, then made a U-turn and returned to the animal's windpipe (figure 16). The sounds may have originated with vocal chords, other vibrating membranes, or air may have been blown directly across the open ends of the bony pipes. Several distinct deep bass frequencies could have been produced by these pipes, and perhaps even a musical chord. Females, with shorter pipes, would produce a higher frequency or pitch. With great lung power and a crest 2 meters long, these dinosaurs could have produced sounds that rumbled for miles across the countryside. A crest of this size would produce a low sound frequency of about 85 cycles per second (Alexander, 1989). This is slightly lower than the tone made by blowing across the top of an empty two-liter pop bottle. There also may have been a bassoonlike sound with rich overtones. Paleontologist Robert Bakker has nicknamed the *Parasaurolophus* the "trombone-duckbill hooter"! If dinosaurs had complex vocal chords, then their sounds may have been even more musical, perhaps comparable to the Australian didgeridoo instrument.

Figure 16

The typical headcrest chamber of a duckbill dinosaur.

There is evidence that some of the long-tailed dinosaurs may have played "crack the whip." Computer simulations show that a 13-meter sauropod tail could have been swung like a giant bullwhip, with the end portions easily reaching the speed of sound. This would cause a miniature sonic boom, heard as a sharp cracking sound. Stress injury has been noticed about a quarter of the way along some fossil tail ligaments. This same position is also a major stress point in a leather whip. Skeptics consider it unlikely that dinosaurs would purposely cause possible pain and injury to themselves by such violent motion. For wooing mates and threatening rivals, however, such loud cracks may have once echoed across the land.

---

50. How good was dinosaur vision?

In past years it was thought that dinosaurs had very poor vision. After all, they were considered prehistoric animals, which must have been simple in nature. *Tyrannosaurus rex,* for example, was thought to lack depth perception. The work of computer scientist Kent Stevens has radically changed this false assumption. Stevens placed glass eyes within the eye sockets of dinosaur models and then measured the range of view. The *T. rex* was found to have about 50 degrees of binocular vision in the forward region where its two eyes overlapped. This overlap permitted excellent stereo or three-dimensional vision for the dinosaur. For comparison, an alligator has only 20 degrees of overlap, while a house cat has about 130 degrees. The optical studies also show that visually *T. rex* did not have to rely on the movement of subjects in order to detect them. Its vision was probably very acute, like many animals today. Another dinosaur skull studied by Stevens, that of the *Carcharodontosaur,* had eyes with very little overlap and therefore less depth perception.

Like animals today, dinosaurs probably displayed great variety in their vision. Each was superbly engineered for a particular lifestyle. The enlarged eyes of many dinosaur species suggest that they could see colors, as do many mammals, birds, crocodiles, and fish today. In the animal world there are more than 40 distinct types of eye structure. Evolutionary theory must assume that all these varieties formed separately, by chance. Instead, however, eyesight declares the Creator's care for His creatures. Psalm 94:9 explains that God formed both ears and eyes.

Conclusion

This book has given a creation overview for the dinosaurs. Three concluding points can be made. First, the dinosaur world does not conflict with a literal, biblical view. Instead, the dinosaurs can be fit readily into biblical history. Bible-science conflicts arise only when earth history is viewed in evolutionary terms.

Second, dinosaur design shows God's glory instead of random evolutionary change. The dinosaurs reveal complex behavior, temperature regulation, unending variety in their appearance, and parental instinct. *All* scientific research is actually creation research, including dinosaur studies. The trouble is that data interpretation often goes astray when not considered in the light of God's Word. Creationists have the privilege of honoring the Creator by acknowledging Him in their studies.

Third, the dinosaurs remind us that even the strong are temporary. As mighty and dominant as they were, these creatures died out and are not present today in our zoos. This lesson should not be overlooked. Each of our own lifetimes is likewise fleeting. It is only what is done for our Creator and Savior that will have lasting value. To know one's Maker personally, as explained in the Bible, is of highest importance in this life and in the life to come.

References

Agosti, D., D. Grimaldi, and J. Carpenter. 1997. "Oldest known ant fossils discovered." *Nature* 391: 447.

Alexander, R. McNeill. 1989. *Dynamics of Dinosaurs and Other Extinct Giants.* New York: Columbia University Press.

Bakker, Robert T. 1986. *The Dinosaur Heresies.* New York: William Morrow and Company, Inc.

Coates, Karen. 1998. "Through Dinosaur Eyes." *Earth* 7, no. 3: 24–31.

Desmond, Adrian. 1975. *The Hot-Blooded Dinosaurs.* London: Blond and Briggs.

Dillow, Joseph. 1981. *The Waters Above.* Chicago: Moody Press.

Dohan, Mary. 1981. *Mr. Roosevelt's Steamboat.* New York: Dodd, Mead and Co.

Farlow, James O. 1981. "Estimates of dinosaur speeds from a new trackway site in Texas." *Nature* 294: 747–48.

Farlow, James O., and M. K. Brett-Surmann, eds. 1997. *The Complete Dinosaur.* Bloomington, Ind.: Indiana University Press.

Fricke, Hans. 1988. "Coelacanths—The fish that time forgot." *National Geographic* 173, no. 6: 824–38.

Horner, John. 1988. *Digging Dinosaurs.* New York: Workman Publishing.

Lauson, Douglas. 1975. "Pterosaur from the late Cretaceous of West Texas: Discovery of the largest flying Creature." *Science* 187: 947–48.

Lockley, Martin. 1991. *Tracking Dinosaurs.* New York: Cambridge University Press.

Mackal, Roy P. 1987. *A Living Dinosaur?* Leiden, Netherlands: Brill.

Matthews, Peter, ed. Annual. *The Guinness Book of Records.* New York: Facts on File.

McGowen, Christopher. 1983. *The Successful Dragons.* Toronto: Samuel Stevens.

References

Monastersky, R. 1997. "Biologists peck at bird-dinosaur link." *Science News* 152, no. 20: 310–11.

Norell, Mark A., James M. Clark, Luis M. Chiappe, and Demberelyin Dashzevey. 1995. "A nesting dinosaur." *Nature* 378: 774–76.

Padian, Kevin. 1986. *The Beginning of the Age of Dinosaurs.* New York: Cambridge University Press.

Schreeve, James. 1997. "Uncovering Patagonia's Lost World." *National Geographic* 192, no. 6: 120–37.

Thomson, Keith. 1991. *Living Fossil—The Story of the Coelacanth.* New York: W. Norton and Co.

Travis, John. 1996. "What dinosaurs left behind." *Science News* 150: 186.

Suggested Resources

Alexander, R. McNeill. 1976. "Estimates of speeds of dinosaurs." *Nature* 261: 129–30.

Calloway, Jack, and Elizabeth Nicholls, eds. 1989. *Ancient Marine Reptiles*. New York: Academic Press. A compilation of technical articles devoted solely to aquatic reptiles.

Dodson, Peter. 1997. *The Horned Dinosaurs*. Princeton, N.J.: Princeton University Press. A definitive work on the *Ceratopsian* dinosaurs.

Eldredge, Niles, and Steven Stanley, eds. 1984. *Living Fossils*. New York: Springer-Verlog. Many living fossils are discussed in technical detail.

Gish, Duane. 1992. *Dinosaurs by Design*. Green Forest, Ariz.: Master Books. This color-illustrated book is written by a leading creation scientist.

Gurney, James. 1992. *Dinotopia*. Atlanta: Turner Publishing, Inc. James Gurney writes books about a mythical land where dinosaurs and people coexist. Dinosaurs are given personalities and high intelligence. Gurney's imagination and illustrations are outstanding.

Ham, Ken. 1998. *The Great Dinosaur Mystery Solved*. Green Forest, Ariz.: Master Books. This is a helpful book by a well-known creationist.

http://www.mtnswest.com/ores/geology/paleo/index.html. This site gives links to all internet paleontology topics.

Milner, Richard. 1990. *The Encyclopedia of Evolution*. New York: Facts on File. The strengths, weaknesses, and history of evolution are clearly written.

Norman, David. 1985. *Illustrated Encyclopedia of Dinosaurs*. New York: Crescent Books. An excellent general book on dinosaurs, although with the usual evolutionary slant.

Shipman, Pat. 1981. *Life History of a Fossil*. Cambridge: Harvard University Press.

Svarney, Thomas, and Patricia Barnes-Svarney. *The Handy Dinosaur Answer Book*. Detroit: Visible Ink Press, 2000. Less dogmatic about evolution than many books.

Suggested Resources

von Fange, Erich A. 1992. *Helping Children Understand Genesis and the Dinosaur*. Published by the author. A very readable dinosaur introduction for children and parents.

Weishompel, David, Peter Dodson, and Halszka Osmolska, eds. 1990. *The Dinosauria*. Berkeley: University of California Press. A technical source for dinosaur studies.

Whitcomb, Norma. *Those Mysterious Dinosaurs*. Published by the author, 1991. A clear description of dinosaurs and also the gospel message.

Creationist Organizations

The following organizations publish catalogs of creationist material, including dinosaur studies.

Access Research Network
P.O. Box 38069
Colorado Springs, CO 80937-8069
Publishes *Origins and Design* journal.

Answers in Genesis
P.O. Box 6330
Florence, KY 41022-6330
Publishes *Creation* magazine and *Technical Journal.*

Apologetics Press, Inc.
230 Landmark Drive
Montgomery, AL 36117-2752
Publishes *Reason and Revelation* journal.

Creation Moments, Inc.
P.O. Box 260
Zimmerman, MN 55398-0260
Publishes *Creation Perspective* magazine.

Creation Research Society
P.O. Box 8263
St. Joseph, MO 64508-8263
Publishes *CRS Quarterly*

Eden Communications
1044 N. Gilbert Rd.
Gilbert, AZ 85234
Produces Christian films and literature.

Institute for Creation Research
P.O. Box 2667
El Cajon, CA 92021-0667
Publishes *Impact* magazine and many creationist books.

New Life Press
P.O. Box 727
Green Forest, AZ 72638
Publishes creation-oriented books.

Glossary

absolute age An actual number of years assigned as the age of an object. Compare with relative age.

amber A hardened clear fossil resin, often yellowish in color. Originally secreted from plants and trees as a viscous liquid.

angiosperm A plant whose seed is surrounded by a closed seed vessel, such as any flowering plant.

ankylosaur Short-limbed, four-legged, armor-plated dinosaurs. In the ornithischian order.

aquatic Living in fresh or salt water.

arboreal Living on land among trees or shrubs.

archosaur The classification group including dinosaurs, birds, pterosaurs, crocodiles, and close relatives. The word means "ruling reptile."

articulated Fossil bones that are found connected or in place, just as they were in the living animal. Scattered bones are disarticulated.

asteroid One of the many "minor planets" or rocky objects that orbit the sun and are located mainly between the orbits of Mars and Jupiter.

avian Relating to birds.

bed A single layer of sedimentary rock.

behemoth A large land-dwelling beast from Old Testament times. The description in Job 40 fits that of *Apatosaurus*.

biomechanics The study of the energy, strength, and motion of living creatures, especially their muscle activity.

biostratigraphy The study of the distribution of fossils in distinct rock layers.

bipedal Using only two limbs for walking, like mankind and birds.

bone bed A sedimentary rock layer containing a large number of fossil bones or fragments.

carbonization A fossilization process in rock layers where only a paper-thin residue of carbon remains from the original organism. Delicate leaves and animal forms are sometimes preserved in this way.

carnivore A flesh-eating animal. Also includes insectivorous plants.

catastrophism The belief that many major physical changes on the earth result from sudden catastrophes rather than gradual evolutionary processes.

ceratopsia Four-legged ornithischian dinosaurs, often with frills and horns.

cladistics The classification of organisms based on supposed evolutionary relationships. Characteristics such as similar bones form the basis for particular groups.

cold-blooded The popular term for ectotherms.

conifer Woody plants including pines, spruces, and cedars. Thought to be the dominant flora of the Mesozoic era, and continuing today.

coprolite Fossilized droppings from any kind of animal.

Cretaceous period The third and last geologic period of the Mesozoic era. Thought to last from about 144 million years ago to 65 million years ago.

cryptozoology The study of rare animals whose very existence is debated.

digit A toe or finger.

dinosaur A collective term that includes the saurischian and ornithischian reptiles.

DNA A complex molecular arrangement within living cells that carries genetic information. DNA consists of two long chains of atoms twisted into an elegant double helix.

ecosystem A local biological community of organisms and their many interactions with each other and with the environment.

ectotherm An animal that obtains most of its body heat from the surroundings. Includes fish, amphibians, and reptiles. Body temperature is usually lower than that of warm-blooded animals.

encephalization quotient (EQ) Estimated brain size, compared with expected brain size for a creature of similar mass.

endotherm A warm-blooded organism that obtains most of its body heat from its own metabolism. Includes birds and mammals; also called homeotherms.

evolution The theory that life began spontaneously on the earth long ago. Over time, living things have supposedly changed on a large, macroscopic scale and given rise to completely new forms.

fauna and flora The animal and plant life of a region.

fossil Preserved remains or traces of a plant or animal, usually embedded in sedimentary rock. From the Latin *fossilis,* meaning "dug up."

gastrolith Stomach stones that help animals grind food after swallowing. Pebbles or small rocks are swallowed by an animal and kept in the digestive tract.

gymnosperm A plant whose seed is not surrounded by a fruit. Includes conifers and seed ferns.

herbivore An animal that feeds chiefly on plants. Includes cows, sheep, most ornithischians, and sauropod dinosaurs.

homology Similarity of organs or body structure between different animals, or between animals and people. Often credited to a common ancestor, homology instead shows the common Creator of all life.

ice age A worldwide cold period with accompanying continental glaciation.

ichnology The study of footprints, teeth marks, and other trace fossils. From the Greek *ichnos,* meaning "trace."

inertial homeotherm An animal that maintains a fairly constant body temperature due to its large size.

invertebrate An animal with no backbone or spinal column. Includes insects and mollusks.

isotope Particular forms of a chemical element, such as oxygen 16, 17, and 18. Isotopes differ in the number of neutrons within the atom's central nucleus.

kind A biblical category of life (Gen. 1:21). The *kind* is similar to the biological class of genus, family, or subfamily.

K-T boundary The separation between rock layers of Cretaceous and Tertiary geologic periods. *K* is commonly used instead of *C* for the Cretaceous. This transition is usually dated at approximately 65 million years ago.

leviathan A large marine reptile from Old Testament times. The description in Job 40 fits that of *Plesiosaurus.*

living fossil A plant or animal that appears unchanged from "ancient" fossils. Most living fossils were once thought to be long extinct.

mammal Any warm-blooded vertebrate creature. Mammals are characterized by skin, hair, and the females having milk-producing glands. People also are classified as mammals.

marine Relating to or living in the ocean.

mass extinction The sudden death of a large number of diverse animal groups.

Mesozoic era The geologic period after the Paleozoic and before the Cenozoic. The Mesozoic is said to have extended from about 208 to 65 million years ago. Also called the Age of Reptiles.

metabolism The total internal activity of organisms including eating, digesting, and making tissue.

mineralization The gradual conversion of a buried plant or animal to hardened inorganic material. Also called fossilization.

morphology The study of the physical form of plants and animals, including their size and shape.

mutation A change in the genetic structure of a plant or animal, often due to exposure to radiation or chemicals.

natural selection The process by which organisms best adapted to their environment tend to survive. Those less suitable are eliminated. Creationists see this as a conservative process that protects the health and stability of created kinds of organisms.

omnivore An organism that eats anything available, whether meat or plant material.

oology The scientific study of eggs, including their fossils.

ornithischia An order of dinosaurs with a pelvic structure somewhat similar to that of birds. The word means "bird-hipped."

ornithopod Means "bird-feet." Describes the bipedal ornithischian dinosaurs, probably herbivores.

osteology The study of bones, including those of dinosaurs.

oviparous Producing eggs that are hatched outside of the maternal body.

paleo A prefix meaning "past" or "ancient."

paleontology The scientific study of plant and animal life from the past, including fossils. Paleontology is often considered a branch of geology.

paleopathology The study of ancient diseases and injuries, often revealed in fossil bone abnormalities.

Panagea (Pangea) The proposed supercontinent that existed when all the continents were joined into one land mass. Later divided into the present continents.

pelvic girdle A connected ring of bones near an animal's hips, supporting the hind legs or fins.

permineralization The addition of minerals to a bone during fossilization. Also called mineralization.

petrification Conversion of organic matter into a hard mineral such as calcite or quartz. The word means "turned into stone."

plate tectonics The principle that the earth's crust is divided into large, rigid plates that float upon the lower mantle. Movement along the plate margins causes earthquakes and volcanoes.

proto A prefix meaning "earliest" or "first."

pseudofossil A natural or man-made object that resembles an actual fossil.

quadrupedal Traveling on four legs.

radiometric dating Measuring amounts of atomic isotopes in rocks or other objects, and then assuming that this data gives the sample age.

raptor A modern bird of prey such as a hawk or eagle. Also includes some dinosaurs that displayed predatory behavior.

relative age The age of one object compared with another, without stating an actual number of years. Compare with absolute age.

reptile A cold-blooded, usually egg-laying vertebrate, with an external covering of scales or horny plates. Examples include snakes, lizards, turtles, and dinosaurs.

saurian Any of the various reptiles including lizards, and in former classifications also crocodiles and dinosaurs. Relating to or resembling a lizard.

saurischia An order of dinosaurs with a pelvic girdle similar to that of modern reptiles. The word means "lizard-hipped." There are two saurischian dinosaur types, the theropods and sauropods.

sauropod A saurischian type of dinosaur that was quadrupedal, a herbivore, and often gigantic in size. They had small skulls, long necks and tails, and massive limbs with five toes.

scavenger An animal that eats the flesh of dead animals without hunting and killing live prey itself.

species A distinct group of organisms having the ability to reproduce readily within the group, producing fertile offspring.

stegosaur Short-limbed, four-legged ornthischian dinosaur with triangular back plates.

taphonomy The study of the conditions and processes by which organisms are fossilized and preserved.

taxonomy The classifying of plants and animals into groups. Schemes are often based on evolutionary assumptions.

terrestrial Living on land as opposed to living in water.

theropod A saurischian type of dinosaur that was bipedal and mainly carnivorous.

trilobite An extinct marine arthropod having a three-lobed body. Very abundant in the fossil record, especially during the Cambrian period.

uniformitarianism The view that geological and biological changes in the past always occurred at the slow rates often measured today. A denial of supernatural intervention in history.

vertebrate Having a backbone or spinal column. Includes fishes, amphibians, reptiles, birds, and mammals including mankind.

warm-blooded The popular term for endotherms.

Dinosaur Name Summary

The following list of dinosaurs and other early animals includes the meaning of scientific names and identifying features. The list is only representative because the total species of dinosaurs alone is nearly 1,000 and rising with each new discovery. The categories include Sauropods (S), Theropods (T), Ornithopods (O), Marine Reptiles (MR), Flying Reptiles (FR), Mammals (M), and Birds.

Name	Meaning	Feature	Category
Albertosaurus	Alberta lizard	Similar to *T. rex*	S
Allosaurus	Different lizard	Large head	T
Anatosaurus	Duck lizard	Duckbill	O
Ankylosaurus	Armored lizard	Bony armor	O
Apatosaurus	Deceptive reptile	Name change from *Brontosaurus*	S
Archaeopteryx	Ancient wing	Modern feathers	Bird
Baryonyx	Heavy claw	Large curved claw	T
Brachiosaurus	Arm lizard	Long forelimbs	S
Brontosaurus	Thunder lizard	Large size	S
Camarasaurus	Chambered lizard	Hollow skull structure	S
Campasaurus	Camp lizard	Beaked jaw	T
Carnosaurus	Meat lizard	Carnivorous	T
Caudipteryx	Tail feather	Ostrichlike	Bird
Ceratosaurus	Horned face	Multiple horns	O
Ceratopsian	Horn lizard	Modest size	T
Coelurosaurus	Hollow-boned	Slender, graceful	T
Compsognathus	Pretty jaw	Very small	T
Corythosaurus	Helmet lizard	Duckbill	O
Cryptoclidus	Hidden tooth	Long neck	MR
Deinonychus	Terrible claw	Sicklelike claw on foot	T
Dilophosaurus	Two-crested reptile	Display crests on nose	T

Diplodocus	Double-beamed	Long, thin	S
Dromiceiomimus	Emu mimic	Very fast runner	O
Elasmosaurus	Ribbon lizard	Long neck	MR
Eoraptor	Dawn stealer	Dog-sized	T
Gigantosaurus	Giant lizard	Largest carnivore	T
Hadrosaurus	Big lizard	Duckbill family	O
Hypsilophondon	High-crested tooth	Moveable joints in skull	O
Ichthyosaurus	Fish lizard	Similar to dolphins	M
Iguanodon	Iguana tooth	Spikes on thumbs	O
Kentrosaurus	Painted lizard	Stegosaur with many spines	O
Kronosaurus	Time reptile	Massive skull	MR
Maiasaurus	God mother lizard	Nests found in Montana	O
Mammoth	Great beast under the earth	Shaggy elephant	M
Megalosaurus	Great lizard	An early find	T
Mononychus	One claw	Turkey-sized	T
Mosasaurus	Sea lizard	Fearsome teeth	MR
Ornitholestes	Bird robber	Ostrich like	O
Ornithomimus	Bird mimic	Birdlike	O
Oviraptor	Egg thief	Crest on head	T
Pachycephalosaurus	Thick-headed	Massive skull bones	O
Phobasuchus	Horror crocodile	Large skull	MR
Plesiosaurus	Nearer to the reptile	Graceful swimmer	MR
Protoceratops	Early horn-face	Horned	O
Psittacosaurus	Parrot lizard	Parrotlike snout	O
Pteranodon	Toothless flier	30-foot wingspan	FR
Pterodactyl	Finger wing	Small, tailless	FR
Pterosaurus	Winged lizard	Leathery wings	FR
Quetzalcoatlus	Flying feathered serpent	Large flying reptile	FR
Rhamphorhynchus	Narrow beak	Long tail and beak	FR
Stegosaurus	Plated roofed lizard	Plates on back	O
Struthiomimus	Ostrich mimic	Like an ostrich	T
Styracosaurus	Spiked lizard	Long spikes on skull	O
Suchomimus	Crocodile mimic	Long narrow snout	T
Titanosaurus	Giant lizard	Large size	S
Triceratops	Three-horned face	Horns on head	O
Tyrannosaurus	Tyrant lizard	6-inch teeth	S
Utahraptor	Birdlike, from Utah	Large claws	T
Velociraptor	Swift robber	Small, swift	T

Scripture Index

Note: References are to question numbers rather than page numbers.

Subject and Name Index

Note: References are to question numbers rather than page numbers.

Don DeYoung chairs the Natural Science Division at Grace College, Winona Lake, Indiana. He is also on the faculty of the Institute for Creation Research, San Diego, California. Dr. DeYoung is currently president of the Creation Research Society. He writes and speaks on creation topics.

Other books by Donald DeYoung:

Astronomy and the Bible
Weather and the Bible
Science and the Bible, I
Science and the Bible, II